PHILCO RADIO

A Pictorial History of the World's Most Popular Radios

1928-1942

by Ron Ramirez
with Michael Prosise

Schiffer Publishing Ltd

77 Lower Valley Road, Atglen, PA 19310

Dedication

To the "Dean" of radio collectors, James A. Fred, who introduced
me to this fascinating hobby through his magazine column in 1974.
And to my wife, Linda, and my daughters, Kandi and Angie, for
bearing with me while I spent many hours preparing this volume.

Designed by Bonnie Hensley

Published by Schiffer Publishing Ltd.
77 Lower Valley Road
Atglen, PA 19310
Please write for a free catalog.
This book may be purchased from the publisher.
Please include $2.95 postage.
Try your bookstore first.

Acknowledgements

The author wishes to extend his most sincere thanks to the following people, without whose help this book would not have been possible.

Ron Boucher, for talking me into covering the years from 1938 to 1942 more in-depth, then graciously loaning me Philco dealer materials to allow me to do so; for providing me with much needed information; and for providing photographs and allowing me to use the great artwork! Your assistance has meant a great deal to me and I thank you.

Carl Bryant, Tim Kaiser, Wayne King and John Miller, for allowing me to use various Philco brochures and catalogs in my research.

John D. Hall, for his help in proofreading the manuscript.

Doug Houston, for supplying numerous photographs, and for his thoughts and advice.

Philips Consumer Electronics Company, for graciously allowing us to use the "Philco" name and many period Philco advertisements in this volume, and for furnishing helpful information.

Michael Prosise, for allowing me to use his text and chart on Philco's 1928 radios in this volume, for supplying numerous pho-tographs and advertisements for this book, for supplying illustrations of Philco sets from 1928, 1930, 1931, 1932, 1940 and 1941, for providing a great deal of research material, and for a tremendous amount of help overall, which I appreciate. Many thanks, Mike.

And, Paul Rosen, for supplying photographs as well as a great deal of information on Philco's Canadian subsidiary.

Sincere thanks also to the following people who also supplied photographs for this volume: Edmund DeCann, Spencer Doggett, Allan Haasken, David Kendall, Wayne King, Jerry McKinney, John Miller, Dave Millward, John Okolowicz, Dennis Osborne, Lewis Owens, Bob Schafbuch, John Sedlacek, and Gary Schieffer.

Also, thanks to Walter Asherbraner, Larry Baehl, Danny Carter, Richard Cason, George Freeman, Roger Hammers, Karl Koogle, James Palmer, Skip Shaw, and Allen Tolson for their help.

Thanks also to Colleen Prosise for her assistance.

And, thanks to Marty Bunis and Dick Mackiewicz for their thoughts and advice.

Finally, a very special "thank you" to Peter B. Schiffer, director of Schiffer Publishing Ltd., for helping me through this, my first book.

Preface

This book came about in response to the lack of pictorial references on Philco radios. It was felt that the antique radio community, as well as anyone else who appreciates antiques and collectibles, could use such a book. Therefore, considerable amounts of time have been spent over the past two years collecting and researching facts and information on Philco, as well as taking and collecting photographs.

From the beginning, I decided color photographs would be best. A color photo gives a person a better idea of how a radio looks than a black and white photo does. More than half of the photographs in this volume are in color. However, in order to have as complete a Philco reference as possible, it was necessary to include many black and white photographs and illustrations as well.

I have grouped the various Philco models into chapters divided by year. Chapters two through five are divided into calendar years. Chapter six covers early 1932, as well as the complete 1932-33 selling season. Chapter seven covers the 1933-34 season. From chapter eight on, the years referred to are model years instead of calendar years.

Every chapter lists each Philco model made during that year with a brief description of each model, and a tube layout (when available). This was done in order to help you identify a particular Philco set made between 1928 and 1942 if you do not know the model number. In the few cases where no photograph was available for a particular model, only a description and (if available) a tube layout is given.

As you will see by looking through this volume, many Philco models shared cabinet styles. In some instances where no photographs are shown for a particular model, you will be referred to another model which looks similar to the one in question. This way, you will have an idea of how the model you are interested in looked.

It was not possible to show every Philco model made during the period covered by this book. If every model were shown, this book would have become an encyclopedia! However, it is felt that the format chosen will suffice as a reference on Philco sets.

I have tried to use the best possible examples of the various Philco sets. However, there are a few photographs included which show sets with incorrect grille cloth, wrong knobs, and/or a scratched or poor finish. While I realize this will be offensive to the purist, it was felt that it would be better to show an example of a set, even if not perfect, than to leave it out entirely. The goal for this book was to show as many Philco radios as possible. However, you will find that the vast majority of sets shown are in excellent or near mint condition.

It was also decided that this book would not have a price guide that would soon be out of date. It was felt that, since this is the first pictorial reference book devoted exclusively to Philco, it should be a stand-alone book that would always be up to date as a reference volume.

The chart at the end of chapter sixteen lists Philco radios sold in Canada between 1928 and 1942, along with their U.S. equivalents. Canadian Philco sets were basically the same as U.S. Philco models until around 1940; by the 1942 season, most Canadian models had become quite different from U.S. Philco sets. The only major difference between the older Canadian sets and their U.S. counterparts is that electric sets sold in Canada were designed to operate on 25 to 40 cycle alternating current (for many years, 25 cycle AC was the standard in Canada), while most U.S. models were designed to operate on 50 to 60 cycle AC.

The appendix at the end of this book lists some of the suppliers of parts and services for old radios, as well as several old radio clubs in the U.S.

A glossary of some radio terms is also included to assist the beginning collector. I tried to make the definitions as simple as possible so that the novice, as well as the non-collector, can understand some of the terms used in this book.

I hope you enjoy looking through this book as much as I enjoyed putting it together. Any comments you may have about this book would be appreciated.

Ron Ramirez.

Contents

When the name "Philco" is mentioned, one immediately thinks of the company's consumer electronics products, most especially the classic "cathedral" radios they were famous for. What you may not be aware of, however, is that Philco's history goes back over 100 years.

Back in 1892, in Philadelphia, Pennsylvania, Thomas Spencer, Frank S. Marr, and three others formed the Spencer Company to make carbon arc lamps. In October of that year, the firm became the Helios Electric Company, at which time Marr became its President.

Helios paid a German concern, also named Helios, to use their manufacturing methods and patents. They operated close to bankruptcy for many years. In fact, things were going so bad for the young company that their plant was closed for two weeks in August 1893 due to a lack of business.

The carbon arc lamp business had come to a halt by 1899, and Helios sold some of their assets to a new firm, the Helios-Upton Company of New Jersey. Then in 1904, the company's directors were told Helios' remaining assets would only bring twenty cents on the dollar at a receiver's sale. Rather than sell at that price, Helios continued on.

It is not known what Helios did between 1899 and 1906, since they had sold their rights to the German Helios patents to the Helios-Upton Company in 1899.

In 1906, Helios reorganized under a new name - the Philadelphia Storage Battery Company. Marr continued as President, while another partner, Edward Davis, became Secretary-Treasurer. The revamped firm was now manufacturing storage batteries, mainly for electric automobiles, trucks, and mine locomotives.

The switch to batteries was a smart move for the company. Although business was not very strong during their first years of manufacturing batteries, it was good enough to allow a move to larger quarters on Emerald Street in 1907. Then in 1909, the growing firm moved to a new location at Ontario and C Streets in Philadelphia. They would remain at this location for many years to follow.

Frank S, Marr, who had been President of the company since October 1892, died on December 1, 1916, and was succeeded by Edward Davis. Meanwhile, the Philadelphia Storage Battery Company was doing quite well. They sold over a million dollars' worth of goods for the first time in 1917.

In 1919, they began an aggressive advertising campaign in many national magazines, which proved to be successful. A large amount of advertising would later become a key factor in Philco's rise to the top of the radio industry. The company's General Manager, James M. Skinner, was responsible for the successful 1919 advertising campaign. Skinner would also play a leading role in their future successes.

The year 1919 was also when the company introduced a new trademark - "Philco."

The early twenties brought along a new curiosity - radio. Many companies were jumping into the manufacture of radios (including even Winchester Repeating Arms and Case, a maker of knives). The Philadelphia Storage Battery Company did not follow this trend, but did begin to offer radio batteries in addition to their line of automobile and truck batteries. By 1923, the company was offering battery chargers so that a person could charge their storage battery at home.

In August, 1925, the company introduced a line of "A-B" battery eliminators. Called "Socket-Power" units, they allowed the owner of a radio to operate the set from the light socket without batteries. These Socket-Power units were Philco's first major sellers; by 1927, nearly a million had been sold.

Philco was becoming more involved in radio by now. They were not only selling radio batteries and Socket-Powers, but had also begun to sponsor the Philco Hour, a weekly national radio broadcast.

By late 1927, the UX-226 and UY-227 AC tubes had been announced, and they were quickly adopted by most radio manufacturers. The new tubes made battery eliminators obsolete, and nearly

wiped out what had become Philco's main business. The company was about to be thrown into a major crisis.

A Philco Socket-Power advertisement from 1926. *Courtesy of Michael Prosise.*

Philco Socket-Power Units

With the exception of simple crystal sets, radio receivers manufactured until the mid to late twenties required the use of batteries. A typical radio used a six-volt storage battery to light the tube filaments; two or three 45-volt "B" batteries (each of which had 22 1/2-volt taps) for the plates of the tubes; and sometimes a 4 1/2 to 22 1/2-volt "C" battery. Most "B" and "C" batteries were the dry type and, therefore, could not be recharged; once they were exhausted, they were replaced. Fortunately, they discharged slowly, especially compared to the six-volt or "A" battery.

Many radio set owners in this time period would often experience the frustration at not being able to listen to the radio because the "A" battery had become discharged! Because of this annoyance, home battery chargers were placed on the market. Philco even offered their own battery chargers.

Another disadvantage of the storage batteries was that, as in modern automobile storage batteries, they contained acid which would occasionally be spilled on the wife's favorite rug when the husband was removing the battery for recharging, or replacing it.

It was natural that a demand was created for something better. That demand began to be met in the mid twenties when several companies began to sell battery eliminators. Some of these were designed to eliminate the "A" battery; some would eliminate only the "B" and "C" batteries; others eliminated all batteries. Many of these new devices used the new Raytheon gas-filled cold cathode rectifiers.

Since radio batteries had become a large part of Philco's business by now, the Philadelphia firm also jumped on the bandwagon in August, 1925, with its own line of Socket-Power battery eliminators.

It is interesting to note that, while more "modern" rectifiers were available at the time, such as Raytheon's cold cathode rectifier tubes and the UX-216 half-wave vacuum rectifier tube, Philco's Socket-Power units used what was really an obsolete technology by then - electrolytic (wet) rectifiers. Philco's Socket-Power units sold well, however; 400,000 were sold in 1926 and 500,000 in 1927.

Philco helped fill a demand for a way to power radio receivers without all the batteries. However, when the Radio Corporation of America (RCA) introduced its new A.C. tubes in late 1927, radio manufacturers rushed to adopt them; which quickly put an end to most radio batteries, and to Philco's Socket-Power business as well.

The following three pages show some of Philco's product offerings in December 1927. *Courtesy of Michael Prosise.*

Stop! Look! Think!
—and you'll get your Philco now

3-Point Superiority

1. **The Famous Diamond-Grid**—the diagonally braced frame of a Philco plate. Built like a bridge. Can't buckle—can't warp—can't short-circuit. Double latticed to lock active material (power-producing chemicals) on the plates. Longer life. Higher efficiency.

2. **The Philco Slotted Rubber Retainer**—a slotted sheet of hard-rubber. Retains the solids on the plates but gives free passage to the current and electrolyte. Prevents plate disintegration. Prolongs battery life 43 per cent.

3. **The Quarter-Sawed Hard-Wood Separator**—made only from giant trees 1300 years old; quarter-sawed to produce alternating hard and soft grains. Hard grains for perfect insulation of plates. Soft grains for perfect circulation of acid and current—quick delivery of power. Another big reason why Philco is the battery for your car.

LOOK FOR THIS SIGN

of Philco Service. Over 5,000 stations—all over the United States. There is one near you. Write for address, if necessary.

Safety demands the strongest, toughest, most powerful battery you can get—a battery that will stand by you in emergencies—that won't expose you to the embarrassments, humiliations, and DANGERS of battery failure.

Thousands upon thousands of car owners today—in record-breaking numbers—are replacing their ordinary batteries with dependable, long-life, *super*-powered Philco Batteries.

They know the Philco Battery—with its tremendous power and staunch, rugged, shock-resisting strength—will whirl the stalled engine—give them quick, sure-fire ignition; *get them off at a touch of the starter.*

The Philco Battery is guaranteed for two years—the longest and strongest guarantee ever placed on a battery of national reputation. But with its famous Diamond-Grid Plates, Slotted-Rubber Retainer, Quarter-Sawed Hardwood Separators and other time-tested features, the Philco Battery long outlasts its two-year guarantee.

Why continue taking chances on ordinary batteries? Why wait for an emergency to show you the absolute need for a dependable, power-packed Philco? Install a Philco NOW and be safe. It will cost you no more than just an ordinary battery.

Philadelphia Storage Battery Co., Philadelphia

The famous Philco Slotted-Retainer Battery is the standard for electric passenger cars and trucks, mine locomotives and other high-powered, heavy-duty battery services.

PHILCO
SLOTTED RETAINER BATTERIES
with the famous shock-resisting Diamond-Grid Plates

A Philco battery ad. *Courtesy of Michael Prosise.*

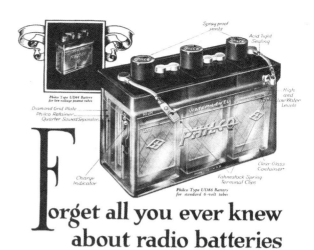

Forget all you ever knew about radio batteries

No more need of big, cumbersome batteries in the cellar for satisfactory, long-distance radio reception.

The new Philco Rechargeable Radio Batteries—assembled in small, attractive, acid-tight GLASS cases—are absolutely safe for use anywhere in your home.

No more need of guess-work charging or using a sloppy old-fashioned hydrometer. The exclusive built-in Philco Charge Indicator tells you all conditions of charge and discharge.

Philco Batteries are Drynamic—shipped DRY charged. Their life starts when YOU pour in the electrolyte—not months earlier at the factory.

Equally important—Philco Batteries deliver a strong, uniform current over long periods. This means great amplifying power—noiseless service—no frequent, troublesome adjustments.

Philco Chargers make recharging so easy, simple and safe a child can do it. Just a throw of a switch—or a plug in a socket. No odor—no noise—no danger of overcharging.

See them at your nearest Philco Service Station, Radio or Music Dealer's, or fill out the coupon below and mail to us.

PRICES

Philco Type UD86 Battery for standard 6-volt tubes. Guaranteed 3 years **$16.00***

Philco Type UD44 Battery for low-voltage peanut tubes.Guaranteed 3 years **$8.00***

* East of the Mississippi River.

The Philadelphia Storage Battery Company, Philadelphia

PHILCO DRYNAMIC RADIO BATTERIES

Another ad for Philco radio batteries, this one from 1924. *Courtesy of Michael Prosise.*

Just pour in the Philco Electrolyte

—and these remarkable new Philco Drynamic Radio "A" and "B" Batteries are ready for use. No waiting for any initial charging. No paying for battery life and current lost before you get them.

Philco Drynamic Radio Batteries are charged DRY at the factory—a revolutionary advance in battery engineering. That means their life doesn't start until the Philco electrolyte is added.

What's equally important—Philco Radio Batteries absolutely free radio reception from "frying," cracking noises in the head phones, and the need of constant "tuning in."

These wonderful new Philco Drynamic Radio Batteries give you the convenience of a dry-cell battery with all the advantages—the absolute dependability and long life—of a Philco Storage Battery, standard for automobiles, mine locomotives and other heavy-duty services.

Equip your radio with Philco Radio "A" and "B" Batteries NOW. You'll be astonished how much better results—at the increased pleasure and satisfaction—you get from your radio set.

Ask your dealer for them or go to the nearest Philadelphia Diamond-Grid Battery Service Station.

RADIO DEALERS—Philco Drynamic Batteries let you into the battery business on a package goods basis. No acid shippage. No charging equipment. No batteries going bad in stock. Wire or write for details.

Philadelphia Storage Battery Company, Philadelphia.

Makers of the famous Philco Slotted-Retainer Batteries—standard for automobiles, electric passenger cars and trucks, mine locomotives and other high-power, heavy-duty battery services.

PHILCO DRYNAMIC RADIO BATTERIES

A 1923 advertisement for Philco radio batteries. *Courtesy of Michael Prosise.*

Early Philco Items

A Philco battery charger from the twenties, with original box and owner's manual.

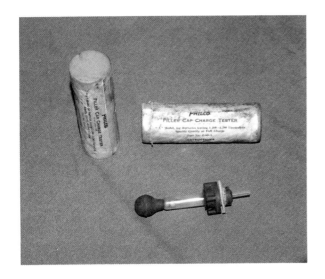

Philco battery tester with original boxes.

Philco Socket Power B, model B-601. *Photo by Michael Prosise.*

Philco Socket-Power A, model A-60.

Philco Socket-Power B, model number unknown, *Photo by Michael Prosise.*

Philco Socket-Power AB, model AB-463.

Philco entered the 1927-28 sales year with grandiose plans for its new 1928 product line, featuring a full line of new Socket Power units, storage batteries and other accessories. At the time, their Socket Power battery eliminator units were their main "bread and butter." In a confidential planning booklet for the year 1928, the company had forecast a sales volume of 1,000,000 Socket Power units, and had set aside an advertising budget of one million dollars to accomplish this goal. It is important to note here that nowhere in this 1928 Planning Guide is there any mention of Philco brand radios.

Still known as the Philadelphia Storage Battery Company, they started out the 1928 season promoting a new feature of the Socket Power units, which they called the "Philco Current Economizer." Advertising for Philco's 1928 products, which officially went on sale in late summer 1927, boasted that "the new 1928 models of the Philco AB Socket Powers are equipped with the new Current Economizer!" In reality this new feature was simply a four-position rotary switch that selectively controlled the ampere charge rate.

Unfortunately for Philco, this Current Economizer, along with all their other fine products of 1928, became nearly worthless months before the actual calendar year began. This major turning point in Philco's history was due to RCA's introduction in September 1927 of an A.C.-powered set (model 17), using RCA's newly developed Alternating Current (A.C.) tubes, the 226 and 227. For the first time, the consumer could simply plug the radio into an electric socket; no batteries required. Philco was notably alarmed, realizing that not only would this severely impact their sales of Socket Powers and radio batteries, but it could eventually put them out of business altogether. The situation soon became even worse for Philco when, late in 1927, RCA announced they would license the use of the new A.C. tubes to other radio manufacturers. Indeed, this spelled the end of the era of battery-powered sets and battery eliminators, as other radio manufacturers quickly seized the opportunity and began making A.C.-powered radios.

Philco now found itself with a product that had become obsolete almost overnight, and, to top it off, the all-important Christmas sales season was just around the corner. However, Philco had no intention of closing its doors and shutting down. A decision was made that they would design and build their own all-electric radios. They hoped to have their first radios available by mid-1928, even though they were already well into the 1927-28 season. Accomplishing this goal would be nothing short of a miracle, requiring long hours and extreme dedication from every employee. Meanwhile, Philco did the best it could with its current, though obsolete, product offering of Socket Powers and accessories. The last Consumer Price Sheet, updating prices for the Christmas sales season, went out to dealers in late November 1927, with all prices and specifications effective December 1. Like the football team that knows they have lost the game, Philco still tried to score a few more points, though in this case it was dollars.

The Legend of the "Neutrodyne-plus"

The quest to get into the production of radio sets began almost immediately. While the legal department scrambled to research various radio patents and licenses, Philco management turned to its engineering department for advice on how they might proceed in finding someone or some company to design a circuit and chassis. As luck would have it, they discovered that one of their main engineers, David P. Earnshaw, was a long-time amateur "ham" radio operator. He knew radio circuitry very well, and had designed and built his own receiving sets along with a few transmitters. Philco therefore decided to place Earnshaw in charge of the engineering aspects of the radio. Management apparently felt confident that Earnshaw could give the company another good product, as he had done when he engineered the successful Socket Power devices.

Throughout the history of the company, Philco had achieved a reputation for quality products, and realized the importance of continuing that tradition in the design of their first radio. If consumers were not impressed with Philco's first sets, they could not be expected to buy another one later. Management therefore placed a high priority on the goal of producing a radio of extreme quality and performance. One of the ways they achieved this objective was to collaborate with the very successful and respected Hazeltine Laboratories of New York, which had designed and introduced the highly regarded Neutrodyne receiver circuit in 1923.

The Neutrodyne design was originally licensed to only a handful of manufacturers, specifically to the 14 original members of the Independent Radio Manufacturers group. In March of 1928, one of those licensees, the Wm. J. Murdock Company, decided to get out of radio receiver production altogether in order to pursue production of other radio-related products. In doing so, they decided to sell all of their RCA and Neutrodyne (Hazeltine) licenses. Philco, having no radio manufacturing rights as yet, was quick to seize upon this opportunity, and negotiated to purchase all the licenses from Murdock for the generous sum of $100,000.

Prior to this, Earnshaw and his staff had already begun to build a prototype chassis based upon the Hazeltine Neutrodyne, endeavoring to improve upon it so they could come up with a Neutrodyne radio that played better than other similar sets currently offered by the competition. A former early employee of Philco, William Denk, states, " . . . they put together a set on the bench, and then took it to Hazeltine [for them] to check over and make suggestions. What they brought back was put into production."

"What they brought back" turned out to be a very good set. It had its own built-in A.C. power supply, employed seven tubes (including rectifier), antenna tuning ("Range Control"), three RF stages, two AF stages, and utilized a large four-gang tuning condenser for "one-knob" tuning, a recent industry advancement from the era of the cumbersome "three-dialer" sets (which required the user to adjust three tuning knobs in order to tune in a radio station). Philco had also installed a phonograph input jack on the front panel, another innovation in those days. In addition, they designed into the circuit the ability of the set to operate without an outside antenna. By placing a "jumper" across the "ANT" terminal post to the "LOC" terminal post, the set utilized the A.C. wiring in the home as an antenna. Until that time, if a person could not or did not want to erect an outdoor antenna, he had to purchase separately a special device that could be screwed into an A.C. light socket, from which he would connect a wire to the radio's antenna terminals. Such devices were common at the time, but Philco was the first to incorporate the concept into the radio itself. It worked very well.

The chassis was designated as model 511 and was used in all of their 1928 radio cabinets. These sets are often referred to as the "511 series," although each cabinet design had its own model number. The chassis was named "Neutrodyne-plus" and Philco pushed the "plus" part heavily in their advertising, claiming that Philco engineers had made a "new radio discovery."

Regarding the word "plus," a former employee says that when he started with Philco in 1937, the story of what the word "plus" meant in "Neutrodyne-plus" had become a much talked about legend at Philco. When questioned about the "plus" many years later, an elderly David Earnshaw recalled that " . . . yes, there was a 'plus', but it wasn't much to get excited about . . ." Essentially, the "plus" was a combination of circuit enhancements and innovations, but the marketing people used them to advertise the set as a "new radio discovery." Specifically, Philco advertised the radio set as having " . . . **super-power . . . perfect tone quality PLUS vast distance range and extraordinary selectivity --a combination new to radio.**"

As best as can be determined by examining Philco's 1928 service manual and through interviews with former early Philco employees, the word "plus" in "Neutrodyne-plus" was only of moderate significance. Philco engineers were able to improve reception of weak, distant stations and improve selectivity (station separation) in various ways, such as with what they referred to as a "range control," operated by a knob mounted on the front right-hand side of the cabinet. When the listener turned this knob, he was actually rotating a small variable tuning condenser. For strong local signals, turning it counter-clockwise would disconnect and ground the grid of the first RF tube, which in turn reduced amplification. This made a slight improvement in tonal quality on very strong signals. When rotated in the opposite direction, fully clockwise, the tube is reconnected in the circuit and enables the Range Control to perform like a "fine tuner" in the antenna circuit.

Selectivity was noticeably better than many other sets on the market, due partly to the use of a large and successfully shielded four-gang tuning condenser.

Philco also touted the fact that their radio receiver would not "howl" or "squeal" when tuning for a station. These loud audible oscillations were a common annoyance of regenerative-type sets, though very few were still in use by 1928. Since radios with "one knob" tuning had only recently become widely available to the consumer, most of the radios in use at the time were still the three-dialer TRF sets, some of which were Neutrodyne "three-dialers". (RCA did have Armstrong's superheterodyne design, but it had not yet been perfected nor was it licensed to any other manufacturers.) Although the TRF type sets were non-regenerative, their tuned RF stages were still prone to cause loud howls and squeals from the speaker. To help counter this problem, the RF stages had to be modified to such an extent that the radio ended up losing much of its sensitivity, which translated into poor reception of distant stations. The Neutrodyne circuit had much better sensitivity, and by its design was not supposed to howl or squeal due to the neutralization process. Unfortunately, some Neutrodyne sets on the market still tended to whistle or squeal occasionally, mostly due to inadequate or absent component shielding, but also because of cheap components and/or inferior circuitry that would lose their neutralization adjustments over time.

Philco believed their sets would never have that problem due to the extensive use of shielding. Philco was the first to shield the RF transformer coils *successfully,* placing them in aluminum cans mounted atop the chassis. Until then, Hazeltine had always insisted it was critical that RF coils be tilted at a very specific angle. Philco's coils, however, were mounted vertically inside the shielding can. The entire set was very well shielded throughout, resulting in very stable and noise-free reception. Even the underside of the chassis was shielded, being completely sealed in by a large metal pan.

According to a few former Philco employees however, some of the sets still tended to oscillate at certain frequencies on the dial, even though they had been properly adjusted and neutralized by standard procedures. It was discovered, by accident, that by purposely misadjusting the first compensating condenser 1/8-to-1/4 turn clockwise, the set would no longer oscillate anywhere across the dial (tuning range). In addition, the engineers were somewhat surprised to find that this also resulted in a further increase of the set's sensitivity and selectivity, most noticeably above 1000 kilohertz. This unusual misadjusting procedure, in conjunction with the range control, the well-constructed four-gang tuning condenser and 100-percent efficient shielding, did in fact make the set perform very well as compared to other sets of that era. However, to call this improvement a "new radio discovery" is stretching the facts a bit. More accurately, Philco's Neutrodyne-plus chassis was simply an extraordinarily well-designed and carefully crafted set. Even today, some 64 years later, a properly restored 511 plays remarkably well.

The engineering of Philco's first radio receiver required the use of 37 different patents--27 from RCA; 10 from Hazeltine, Latour and others. The oldest patent dates all the way back to 1913 while the most recent is 1927.

Unlike the chassis, the Philco speaker of 1928 was not a Philco-engineered item. The speaker they used was obtained by the outright purchase of a small radio products manufacturing company owned by Mr. John S. Timmons. Located in an old converted mill near Philadelphia--in Germantown, Pennsylvania--Timmons Radio Products had been one of the first to manufacture battery eliminators, similar to Philco's Socket Power units. Therefore, like Philco, Timmons now also found one of its main products obsolete. Fortunately for Timmons, their speaker was not, thanks to an oversight by Philco. The Timmons speaker was of magnetic design, which was inferior to the recently introduced electrodynamic-type loudspeaker. Technically, the magnetic-type loudspeaker was indeed obsolete. Had Philco realized this at the time it is doubtful they would have bought the Timmons company; and had this happened, the "Neutrodyne-plus" would have ended up with an electrodynamic speaker that would have greatly improved its tonal quality, thereby making the word "plus" very significant. The well-designed "Neutrodyne-plus," if coupled with an electrodynamic speaker in the year 1928, would have made Philco's first radio sets true state-of-the-art receivers! Unfortunately, this would not happen until 1929.

After the technical aspects of the radio had been put into motion, Philco turned its attention to the marketing aspects. They would need to come up with a selection of designs for the radio cabinets. Since they were entering the radio market late in the season, they wanted something different, something eye-catching, something that would attract the potential buyer immediately. What they came up with was truly brilliant: Flowers!

Greetings, Mademoiselle Messaros!

While the cold winter winds of January 1928 whistled past the Philco offices at Ontario & C streets, a group of Philco planners were busy inside discussing various cabinet designs and how many different styles to offer. Originally it was planned that Philco would introduce four radio models for 1928--a highboy console (model 551/561), a lowboy console (model 531/541), a highboy console with a built-in phonograph (model 571/581), and a metal cabinet table-top (model 511/521) with a separate mantel-type speaker or an optional wooden console speaker cabinet (model 221), upon which the table radio would sit. All the cabinets had an identical chassis, although each cabinet style was also available with a slightly modified circuit in the chassis to accommodate geographical areas using 25, 30, or 40 cycles A.C. instead of the usual 60 cycles. The "other cycle" versions were assigned a different model number (indicated here by use of a slash mark).

For the furniture models, Philco hired an internationally known furniture designer by the name of Albert Carl Mowitz. He submitted several designs for the wooden consoles, and Philco settled on what is described in their literature as "a modern conception of the Louis XVI period style." These cabinets featured walnut panels, a pull-out drawer, doors, fluted legs and fluted pillars. The metal cabinet table model was an attractive design created by a gentleman named Hollingsworth Pearce, reported to have been a leading authority on cabinet design at the time. Philco documents from 1928 show that his design was accepted on March 10, 1928, and was designated as the model 511/521. Its finish was a two-toned Spanish Brown with gold pin-stripes. The top or "lid" of the cabinet was not attached and could simply be lifted off. Philco placed an initial order for 47,350 of these cabinets with American Metal on April 13, 1928, at a cost of $1.14 each, and later ordered an additional 18,500 from Hale and Kilburn. A Philco specification chart shows a total run of 68,850 metal cabinets.

At about the time Pearce submitted his cabinet design for the model 511, management had already begun to consider the idea of offering the 511 in more than one color. It was thought that brightly colored cabinets would naturally stand out, especially among the typical offerings of other radio makers with their mostly brown, black or gray cabinets. The colorful sets would help attract attention to the fact that Philco was now a manufacturer of radio sets, and no longer "the 'socket power' company."

A search was begun to find the right person for the job. Among several different artistic associations in Philadelphia, the name Matild Messaros was often mentioned. At the time, Miss Messaros was a free-lance artist who had been creating beautiful hand-decorated designs on lampshades, furniture, and other in-home furnishings. She was also a graduate of the old Philadelphia School of Design for Women. Philco decided she was just the person they were looking for.

As Miss Messaros recounted during an interview in 1977, "Philco had just started production of its first radio receivers when they contacted me in March of '28. They asked me if I could hire and supervise several dozen artists to hand-paint their metal cabinet radios as they came off the production line." She answered yes; and, at an excellent salary of fifty dollars a week, began a project that produced four beautifully decorated models that today are highly prized by collectors.

Documents from Philco claim that "hundreds of sample cabinets were finished and submitted" during the design process. Eventually, they settled upon four of Miss Messaros' different floral designs, which were applied over each model's base color. In addition to the undecorated two-tone Spanish Brown model 511/521, there would now also be the models 512/522, 513/523, 514/524, and 515/525. The model 512/522 was finished in a bright two-tone Mandarin Red with a floral design of black and gold; while the model 513/523 was a two-tone black and gray finish, called Labrador Gray, with white daisies on gold leaves. The model 514/524 had yellow, tan and gold flowers with black leaves over a dark two-tone Nile Green finish. Finally, the model 515/525, called Impressionistic and looking very different from the others, was painted gold. Its design consisted of green, red and blue lotus leaves rising up from the base with wide green lines outlining the side, rear and front panels, and a continuous red line all the way around the base. This set is the rarest of the group, and very possibly the rarest of all Philco radios as it appears that very few were sold. The hand-decorated "flower radios" were painted only in response to orders from dealers.

The process of painting and applying the floral designs by hand was very time-consuming. The cabinets arrived with a primer coat, over which the base color was sprayed. After it dried, the lighter shade of the same color was applied to give the blended two-tone effect. The cabinet then went to the artist. Philco had hired about 25 artists, all women. They used a stencil to create the basic outline of the floral pattern. Using oils, the artist then created the actual design, all by free-hand. Each cabinet was therefore slightly different. The gold striping was also applied by hand, using a small-tipped brush. The final step after all the designs were in-place and dry, was to spray the entire cabinet with a clear coat of lacquer.

Painting the floral design on the lid alone took ten minutes, while completion of all the floral designs for a particular cabinet could require anywhere from 30 minutes to an hour. This labor-intensive process went on throughout the nine months that these sets were in production. An exact accounting of how many of each style were sold is not known at this time, although it appears that the 511/521 Spanish Brown was the best-selling model in 1928. At $115, it was also the least expensive. The four hand-decorated models sold for $125 each. The prices quoted here, taken from the May 1, 1928, Dealer Price sheet, did not include tubes or the speaker.

The hand-decorated sets also had color-coordinated mantel speakers. Painted with the same base color as the matching radio cabinets, they each had a small floral design on the base with gold pin stripes along the edges. In addition, the decorative mantel speakers had their own model numbers. The Mandarin Red speaker was designated model 212, while the Labrador Gray speaker was listed as model 213. Speaker model 214 had the Nile Green finish, and the gold Impressionistic speaker was model 215. As regards the pricing of the speakers, the plain Spanish Brown model 211 was listed at $25, while the hand-decorated speakers were $27.50 each. The floor model speaker, standing just under 30 inches tall, was called the Console Grand Speaker and was assigned model number 221. It sold for $50, and was quite attractive with one of the 59-pound metal sets sitting on top of it.

One day in early spring, after the artists were in-place and production was finally under way, Miss Messaros was approached by someone from the marketing department. She stated that this gentleman wanted to know if she would object to being referred to as "Mademoiselle" Messaros in Philco's advertising of the sets. She told him it did not make much difference what they called her. Shortly thereafter, Philco's first ads began appearing, referring to the hand-decorated sets as "**. . . enhanced with color effects by Mlle. Messaros, one of the foremost colorists in the decorative arts.**" Probably written by the same "ad man" who called the sets 'a new radio discovery,' the colorful advertising was spectacular, and drew a lot of public attention to Philco's first radios.

Oh What A Year It Was . . .

As reported at the time in the June '28 edition of *Radio Retailing,* a trade journal, Philco unveiled their first line of radio receivers at an industry trade show early in June. The Philco line-up consisted of eight models, the three consoles and the five table-top models, as described earlier. This was quite a landmark occasion for the Philadelphia Storage Battery Company, a company that many thought would become a permanent casualty of RCA's A.C. tubes. In the span of only seven or eight months, a small company that once made "socket power" battery eliminators, was now full steam into the radio manufacturing business, bringing forth an impressively performing high-quality set with cabinets that were well designed and very attractive.

Under the dynamic leadership of General Manager James M. Skinner, Philco had now become a force to be reckoned with. As mentioned, a series of costly double-page ads, in full color, began appearing in every major magazine, emblazoned with the words "**NEW RADIO DISCOVERY! NEUTRODYNE PLUS**" and "**COLOR! VIVID COLOR.**" The ads always featured the beautiful, hand-decorated "flower" sets, and hailed their "super-power" long distance reception. Along with the magazine advertising, Philco also had an established network of newspaper advertising in no less than 663 cities! In addition to all this, there was the weekly broadcast of the Philco Hour every Friday night, heard nationwide over 26 stations. In a promotional pitch to their dealers, Philco pointed out that they " . . . have arranged a series of announcements during the Philco Hour whereby the word Philco is worked into it a dozen or more times during the broadcast."

Philco's aggressive advertising campaign, along with their huge network of dealers, paid off, as sales for 1928 totaled 96,000 units. Although this figure placed them at only 26th in the industry, it was still an impressive showing considering the fact that they were unable to introduce their radios until June. In terms of dollars, Philco managed to generate $12,500,000 in total sales for the year.

The birth of the Philco radio was a significant event in radio history. At a time when radio was just getting out of its infancy, along comes a user-friendly set that established new standards in quality and creative appearance. But more significant was the birth of Philco as a new manufacturer of radios, because in only two more years, Philco would take the number one position in the industry, selling twice as many radios as its nearest competitor. This is a position the company would maintain for many years to come, and along the way would introduce many innovative and exceptional products.

As the year 1928 drew to a close and the last flower was brushed upon the last set, Philco's 1929 line-up of new radios was already waiting in the wings. There would be no more flower radios next year, but the legendary Neutrodyne-plus would carry on, and this time would be even better than before.

Philadelphia Storage Battery Company, Philadelphia, Pa

Ontario and C Streets, Phone--REGent 8840

Philco Radios for 1928

Model Number	110-Volt A.C. Cycles	Finish	Consumer Prices (Set without Tubes)
Highboy Console			
551	50/60	Walnut	$275.00
561	25/30/40	Walnut	285.00
Deluxe Electric Phonograph-Radio Console			
571	50/60	Walnut-Maple	not available
581	25/30/40	Walnut-Maple	not available
Lowboy Console			
531	50/60	Walnut	200.00
541	25/30/40	Walnut	210.00
Table Models			
511	50/60	Spanish Brown	115.00
521	25/30/40	Spanish Brown	125.00
Table Models--Hand-Decorated			
512	50/60	Mandarin Red	125.00
513	50/60	Labrador Gray	125.00
514	50/60	Nile Green	125.00
515	50/60	Impressionistic	125.00
522	25/30/40	Mandarin Red	135.00
523	25/30/40	Labrador Gray	135.00
524	25/30/40	Nile Green	135.00
525	25/30/40	Impressionistic	135.00

Philco Speakers

Console Grand Speaker

221		Walnut	50.00

Mantel Type Speakers

211		Spanish Brown	25.00

Mantel Type Speakers--Hand-Decorated

212		Mandarin Red	27.50
213		Labrador Gray	27.50
214		Nile Green	27.50
215		Impressionistic	27.50

Bibliography

Colliers, October 13, 1928.

Denk, William, former Philco employee, interview with author, April 28, 1991.

Denk, William, interview with author, June 20, 1991.

Denk William, and Denk, Jane, "Hand Painted Philcos," *ARCA Gazette,* Vol.5, No. 3, 1977.

Douglas, Alan, *Radio Manufacturers of the 1920's,* Vol. 1, New York, New York, Vestal Press, LTD., 1988.

Douglas, Alan, *Radio Manufacturers of the 1920's,* Vol. 2, New York, New York, Vestal Press, LTD., 1989.

Douglas, Alan, *Radio Manufacturers of the 1920's,* Vol. 3, New York, New York, Vestal Press, LTD., 1991.

Fortune, "1,250,000 Out of 4,200,000 U.S. Radios," Vol. XI, No. 2. New York, Time, Inc., 1935.

Fortune, "Radio, Refrigeration And Radar," Vol. XX, No. 11, New York, Time, Inc., 1944.

Grinder, Robert E., and Fathauer, George H., *The Radio Collector's Directory and Price Guide,* Scottsdale, Arizona, Ironweed Press, 1986.

Husted, Robert, former Timmons and Philco employee, interview with author, April 27, 1991.

Liberty, September 15, 1928.

Lyon, Edwin, telephone conversation with author, July 1992.

Lyon, Edwin, "The Radio Licensing Game, 1921 to 1931," *MAARC Newsletter,* Laurel, Maryland, Vol. 3, No. 3, 1986.

Lyon, Edwin, "The Neutrodyne Mystique," *MAARC Newsletter,* Laurel, Maryland, Vol. 3, No. 7, 1986.

Moyer, James A., and Wostrel, John F., *Practical Radio Construction and Repairing,* New York, New York, McGraw-Hill Book Co., Inc., 1927.

Patterson, Donald, "Philco In The 30's," *Radio Age,* Augusta, Georgia, Patterson Publishing, Vol. 6, No. 6, 1980.

Philadelphia Storage Battery Company, Pamphlet: "Dealer Prices and Information," Philadelphia, Pennsylvania, May 1928.

Philadelphia Storage Battery Company, Brochure: "Neutrodyne Plus - A New Triumph in Radio," Philadelphia, Pennsylvania, 1928.

Philadelphia Storage Battery Company, *Plans for 1927-1928,* Philadelphia, Pennsylvania, 1927.

Philadelphia Storage Battery Company, *Philco Radio Manual of Useful Information 1928-1929,* Philadelphia, Pennsylvania, 1928.

Philadelphia Storage Battery Company, *Radio Installation and Service Manual 1927-1928,* Philadelphia, Pennsylvania, 1928.

Philco Corporation, "Philco Corporation and 60 Years of Progress, 1892-1952," *Philco Hi-Hat Club Newsletter,* Philadelphia, Pennsylvania, Vol. 5, No. 4, July 1952.

Philco Corporation, *Philco Furniture History,* 1928-1940, Philadelphia, Pennsylvania, 1940.

Radio Retailing, June 1928.

Stokes, John W., *The Golden Age of Radio In The Home,* New Zealand, Craig, 1986.

The Saturday Evening Post, November 7, 1925.

Most of Philco's 1928 radio line is featured in this two-page advertisement of the period. *Courtesy of Michael Prosise.*

Models 511, 512, 513, 514, 515, 531, 551 & 571
(all used the same chassis)
Circuit: Neutrodyne-Plus
Frequency Coverage: 550-1500 kc
Power: AC; 115 volts, 50-60 cycles
Tubes Used: 7
Controls: 4; Tuning, Volume, Off-On, Range
Variations: Models 521, 522, 523, 524, 525, 541, 561 & 581 -identical to models 511, 512, 513, 514, 515, 531, 551 & 571, respectively, but for 115 volts AC, 25-40 cycles
See Appendix II (2-1) for the tube diagram for Models 511, 512, 513, 514, 515, 531, 551 & 571

Model 512 in Mandarin Red with matching model 212 speaker. *Photo by Michael Prosise.*

Model 511 in Spanish Brown with matching model 211 speaker. *Photo by Michael Prosise.*

Model 513 in Labrador Grey with matching model 213 speaker. *Photo by Michael Prosise.*

Model 531 console. *Photo by Michael Prosise.*

Model 514 in Nile Green with matching model 214 speaker. *Photo by Michael Prosise.*

Model 515, the "Impressionistic." *Photo by Michael Prosise.*

The De Luxe
Electric Phonograph Radio
Including Console Grand Speaker

This Louis XVI cabinet is constructed with
infinite care of beautifully matched walnut
panels, inlaid with selected Bird's-Eye Maple.

Sizes: 47 inches high, 32 inches wide,
18 inches deep

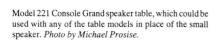

Model 571 Radio-Phonograph. *Courtesy of Michael Prosise.*

Model 221 Console Grand speaker table, which could be used with any of the table models in place of the small speaker. *Photo by Michael Prosise.*

Model 551 is in a cabinet similar to model 86 highboy, shown in Chapter Three on page 20.

Philco had finished 1928 in twenty-sixth place among radio manufacturers, not a bad showing for a company that had been making radios for less than a year. However, another firm that began to manufacture radios in 1928 - the Grigsby-Grunow Company of Chicago, makers of Majestic radios - had done much better.

While Philco was taking its time, building a good reputation for quality, Majestic was seriously challenging the radio leader, Atwater Kent. Nevertheless, Philco's aim for quality would prove to be the correct path, as it ultimately helped the company to grow.

The main reason Majestic did so well in 1928 was the sound of their sets. Majestic receivers were heavy on bass response, unlike any radio had sounded before. Most radios still sounded "tinny," somewhat like a telephone. Some manufacturers, notably RCA, even put tone filters between the audio output and the speaker to cut off low frequency response in order to eliminate any possibility of AC hum. The end result was a tinny sound. (High fidelity reproduction was still many years in the future.) Majestic's tone was largely due to the new electrodynamic speakers used in their sets. Philco, on the other hand, had used high-impedance magnetic speakers which had limited frequency response and were nearing obsolescence.

Although cone-type magnetic speakers had been around for only a couple of years, it was typical of the rapid advancements in radio technology during that time; a design that might be considered advanced one year could very well be obsolete the next.

Philco answered Majestic in January 1929 with the introduction of a new Neutrodyne-Plus, model 86. It was very similar to the 511 series of 1928, with two major differences; the 86 had a more powerful push-pull output stage using two 71A tubes, and an electrodynamic speaker instead of the earlier magnetic speaker. Model 86 was not offered as a table model; it was available only in console cabinets similar to the earlier models 531 and 551. Its sound was not boomy like Majestic, but it wasn't tinny either. Philco pointed out the tone quality of its sets in their advertising, and began to refer to their receivers as "Balanced-Unit Radio."

Philco's next move was to try lowering prices to gain greater sales volume. To accomplish this, the company decided to adopt assembly line techniques. (During this time, most radio manufacturers put together each set by hand, one at a time.) Philco borrowed seven million dollars to expand its plant and convert to mass production.

Meanwhile, Philco introduced two new receivers in June, models 65 and 87. Model 65 was offered in a metal table model cabinet, (similar to the previous year's model 511) that had been designed by Hollingsworth Pearce. However, this year it was available only in two-tone Spanish Brown, with no hand-painted decorations. Both the 65 and 87 were available in lowboy cabinets (which were designed by David Roberts), as well as highboys with double doors and "Deluxe" highboys with sliding doors.

Models 65 and 87 were among the first radios to incorporate tuning dials calibrated in kilocycles, with the last digit removed; i.e. 55 on the new Philco dial meant 550 kilocycles. The dials on most other radios were still calibrated with an arbitrary 0 to 100 scale. Some manufacturers were still using 0 to 100 dials by the early thirties.

The new 65 contained two advanced features. It used two of the new UY-224 screen grid RF amplifier tubes, and two of the new UX-245 audio output tubes. The 224 tubes were better amplifiers than the UX-226 tubes, and they did not require neutralization. The 245 tubes offered greater power output (around four watts push-pull).

The 87, which was another Neutrodyne-Plus model, was nearly the same as the 86, with control knobs closer together and a rotary off-on switch instead of the previous toggle switch. It also used two of the new 245 power tubes.

In 1929, Philco sponsored its first broadcast of the Philadelphia Orchestra, led by well-known conductor Leopold Stokowski. The broadcast aired over the National Broadcasting Company or NBC, which was owned by one of Philco's chief competitors - RCA!

Further evidence that radio was changing rapidly in this time period can be seen by the fact that Philco introduced three new sets in October which were superior to the models they replaced. Model 40 was Philco's first set designed to operate from direct current (DC) mains, which were still present in some locations. It used six tubes in a circuit nearly identical to the new model 76, a seven tube screen grid set which operated on AC.

However, the set that really offered some very good technological advances was the new model 95 Screen Grid Plus. It had the screen grid 224 RF amplifiers and push-pull 245 power output. It was also one of the first radios to use automatic volume control (AVC), a circuit which keeps radio signals at a constant level and prevents blasting of local stations when tuning, and also counteracts fading of distant stations. With the new AVC circuit, one no longer had to keep their hand on the volume control while tuning in a broadcast.

The three new sets were available in the same lowboy, highboy, and "Deluxe" highboy cabinets that had been used for models 65 and 87. In addition, models 76 and 95 were available in the two-tone brown metal table model cabinets, and models 40 and 76 could also be purchased in a new "console" cabinet, a small wood case with long legs.

On October 29th, the stock market crashed, plunging America into the Great Depression. Philco had only been making radios for just over a year, and the company had gone deep into debt to convert to mass production. Would they be able to survive?

An ad for the Screen Grid Plus model 95 from early 1930.

The Screen Grid Plus

Philco's model 95 "Screen Grid Plus" was introduced in October 1929. Philco used the word "Plus" to indicate its best receivers through 1931. Up until now, the "Plus" was more or less a sales pitch as the earlier Neutrodyne-Plus sets had barely kept up with then-modern radio technology.

But in this new set, the "Plus" really meant something special. Its design was far in advance of most of the rest of the industry. The 95 used nine tubes. None of the UX-226 triode tubes were used in this set. (Although the 226 tube was only two years old, it was already nearing obsolescence!) It had three UY-224 screen grid tubes for the best possible RF amplification at the time; three UY-227 triodes, one of which was connected as a diode detector with the other two used as audio amplifiers; and two of the "powerful" (approximately four watts push-pull) UX-245 audio output tubes. In short, all the latest advances in radio.

The 224 screen grid tube was a significant advancement in tube technology, and was quickly adopted by many manufacturers for use in their sets. The new Philco, however, deserved the name "Plus" due to its automatic volume control (AVC) circuit. RCA, the nation's largest radio company, had developed an AVC circuit in 1928, but it required the use of an extra tube. Philco's AVC circuit, which was developed by Harold Wheeler of the Hazeltine Laboratories, did not need an extra tube; the detector tube took care of detecting the radio signals and developing a negative AVC voltage, which was fed back to the control grids of the RF amplifier tubes to control their amplification. The result was that when a signal was tuned in, a negative voltage was developed which depended on the strength of the signal; if it was strong, the voltage was more negative, which reduced the RF amplification. If the signal was weak, the voltage was less negative, and the RF amplification was higher. This prevented strong local stations from coming in at high volume, thereby simplifying tuning and avoiding giving the listener nerve shock.

The standard UX-280 rectifier tube was used in the power supply. Many other manufacturers were still building power supplies for their sets as a separate chassis, connected to the main chassis by a multi-conductor cable. However, Philco's sets had used power supplies built into the main chassis from the very first 511 back in 1928.

The 95 was also one of the first sets to have the volume control in its audio section. Most radios of the time still had volume controls in the antenna circuit, which was not always effective on strong local stations.

Of course, the 95 used an electrodynamic speaker. It also had two tuned circuits ahead of the first RF amplifier tube, which increased selectivity. It also had a local-distance switch. Models 40 and 76 also had these features.

With all it had to offer, it is easy to see why Philco called their new model 95 "Screen Grid Plus."

Models 40, 76 and 95 were on the market from October 1929 through May 1930.

Model 95
Circuit: Screen Grid Plus TRF
Frequency Coverage: 550-1500 kc
Power: AC; 115 volts, 50-60 cycles
Tubes Used: 9
Controls: 4; Tuning, Volume, Off-On, Local-Distance
Variation: Model 92 - for 115 volts AC, 25-40 cycles
See Appendix II (3-6) for the tube diagram for Model 95.

Model 95 was available in cabinets identical to Model 76 table model, lowboy, highboy and Deluxe highboy, shown on pages 19 and 20.

Model 40
Circuit: Screen Grid TRF
Frequency Coverage: 550-1500 kc
Power: 115 volts DC
Tubes Used: 6
Controls: 4; Tuning, Volume, Off-On, Local-Distance
See Appendix II (3-1) for the tube diagram of Model 40.

Model 40 was available in cabinets identical to Model 76 console, lowboy, highboy and deluxe highboy, shown on pages 19 and 20.

Model 65
Circuit: Screen Grid TRF
Frequency Coverage: 550-1500 kc
Power: AC; 115 volts, 50-60 cycles
Tubes Used: 6
Controls: 4; Tuning, Volume, Off-On, Local-Distance
Note: Early production models only had three controls with no local-distance switch
Variation: Model 62 - for 115 volts AC, 25-40 cycles
See Appendix II (3-2) for the tube diagram for Model 65.

Model 65 lowboy console. *Photo by Edmund DeCann.*

Model 65 was also available in cabinets identical to Model 76 table model, Highboy, and Deluxe Highboy, shown below and on pages 19 and 20.

A backlit Philco sign of the period. *Photo by Michael Prosise.*

Model 76
Circuit: Screen Grid TRF
Frequency Coverage: 550-1500 kc
Power: AC; 115 volts, 50-60 cycles
Tubes Used: 7
Controls: 4; Tuning, Volume, Off-On, Local-Distance
Variation: Model 73 - for 115 volts AC, 25-40 cycles
See Appendix II (3-3) for the tube diagram for Model 76.

BALANCED UNIT TABLE MODEL

The *Philco Screen Grid* is available in an attractive table model, complete with genuine *Electro-Dynamic Mantel Speaker.*

Price, Screen Grid Receiver $67.00

The Table Model.

Model 76 console. *Photo by Michael Prosise.*

Model 73 lowboy. *Photo by Allan Haasken.*

The Highboy.

19

BALANCED UNIT HIGHBOY DE LUXE

The Deluxe Highboy.

Closed (left) and open (right) views of Model 86 highboy. *Photos by Wayne King.*

Model 86
Circuit: Neutrodyne-Plus
Frequency Coverage: 550-1500 kc
Power: AC; 115 volts, 50-60 cycles
Tubes Used: 8
Controls: 4; Tuning, Volume, Off-On, Range
Variation: Model 82 - for 115 volts AC, 25-40 cycles
See Appendix II (3-4) for the tube diagram for Model 86.

Model 87
Circuit: Neutrodyne-Plus
Frequency Coverage: 550-1500 kc
Power: AC; 115 volts, 50-60 cycles
Tubes Used: 8
Controls: 4; Tuning, Volume, Off-On, Range
Variation: Model 83 - for 115 volts AC, 25-40 cycles
See Appendix II (3-5) for the tube diagram for Model 87.

Model 87 highboy. *Photo by Michael Prosise.*

Model 87 was also available in a cabinet identical to Model 76 Deluxe highboy, shown on page 20.

Model 86 console.

Model 87 lowboy.

Chapter Four: 1930

By now, Philco had jumped to third place in the radio industry, having sold 408,000 sets in 1929. Despite the fact that America was entering the Great Depression, Philco was doing well. They paid off their entire seven million dollar debt early in 1930.

Because of the stock market crash, most manufacturers expected sales to be lower in 1930. Many even reduced their amount of advertising. Meanwhile, Philco increased its advertising as the company wanted to achieve the number one spot in the industry.

Philco put great emphasis on sales. As an example, at a meeting of Philco dealers during this time, a banner was hung from the ceiling which read "Q (quota) or Q (quit)."

In the spring of 1930, while models 40, 76 and 95 were still being sold, Philco began a promotional tie-in with Paramount Pictures in which a particular Paramount release would also feature Philco Radio in its billboard advertising. Philco arranged for its dealers to hold promotions such as parades, contests, giveaways of autographed pictures of Paramount stars, and even giveaways of Philco radios. Philco dealers also set up displays of Philco sets in theatre lobbies. It was part of Philco's push for more sales.

In June, Philco's new 1930 lineup was introduced. The big news from Philco this season was a four-position tone control, available on every new model. Philco also introduced the Illuminated Station Recording Dial; translucent, back-lit dials which had space to write favorite stations on them with a pencil.

The new Philco models included the 41, 77 and 96. These used chasses which were nearly identical to the previous models 40, 76 and 95. The three now used the new tone controls instead of local-distance switches. Philco also introduced its first battery radio in June. The Battery-Operated Screen Grid Plus model 30 used eight tubes and had many of the features found in the AC-operated Screen Grid Plus model 96, such as AVC, tone control, and push-pull audio output. However, its Station Recording dial did not light up.

The new models had two new cabinets to go with them, a highboy and a lowboy. In addition, models 77 and 96 were available in metal table model cabinets, and models 41 and 77 were also available in the "console" cabinet.

Philco also introduced a new radio-phonograph console. Model 296 used the Screen Grid Plus 96 chassis with an electric phonograph, housed in a large lowboy cabinet. This new cabinet, along with Philco's new highboy, was designed by Edward L. Combs, who would design radio's most famous cabinet less than a year later.

In August, a new top end model was introduced. The Concert Grand also combined the Screen Grid Plus 96 chassis with an electric phonograph. It was housed in a large, elaborate corner cabinet with double doors and sold for $350, less tubes. Philco called it the "greatest of all receivers."

However, it would not be the Concert Grand, Screen Grid Plus, Station Recording Dials or Tone Control that would make big news for Philco in 1930 - it was another model introduced in August that would make Philco a household word.

The model 20 or Baby Grand was not the first cathedral radio ever made, but Philco was the first of the major manufacturers to offer a "midget" set. And the selling price ($49.50 less tubes) made it an instant success, even though it did not have Tone Control. It is no wonder that more than half of the radios Philco sold in 1930 were Model 20 sets!

The new Philco radio line also included another new feature, Philco brand tubes. However, Philco did not make these tubes themselves; they were made by Sylvania for Philco. Besides using Philco tubes in their own sets, the company also offered a line of these tubes for replacement purposes.

Bakelite block condensers, which are small bakelite containers which have condensers and sometimes resistors sealed inside, made their first appearance in the new Philco sets in June. The company was beginning to manufacture most of the parts that went into their sets themselves. Prior to this, Philco had used condensers in its sets that were purchased from outside sources, such as the Aerovox Company.

All models except model 20 continued to use the large, heavy chassis that Philco had been using since 1928. The chassis used in model 20, however, was a hint of future Philco chassis design. It was smaller and lighter than any other Philco model, yet allowed seven tubes and all necessary components to be mounted in it without overcrowding.

1930 was the year Philco formed its own symphony orchestra, conducted by Howard Barlow. Philco also had its own "concert orchestra" with Billy Artzt at the baton, as well as the "famous Philco Quartette."

Philco sponsored three more broadcasts of the Philadelphia Orchestra, conducted by "the world's greatest conductor," Leopold Stokowski, in 1930 over NBC. In addition, Philco sponsored weekly broadcasts of its own symphony orchestra Tuesday nights on CBS.

In the fall, Philco continued its promotional tie-in with Paramount Pictures. A different Paramount movie was featured each month in Philco's national magazine advertising. In September, Philco ads featured Charles "Buddy" Rogers in *Follow Thru,* in October, comedian Harold Lloyd in *Feet First,* in November, Mitzi Green in *Tom Sawyer,* and in December, Ruth Chatterton in *The Right To Love.* Each month during the promotion, Philco dealers could put up billboards which advertised the feature Paramount movie for the month as well as the local Philco dealer. In addition, parades, contests and theatre lobby displays were held by Philco dealers as had been done in the spring.

By October, a new Baby Grand Console had been added to the Philco line. It was actually a consolette, standing only about thirty-three inches tall and used the same model 20 chassis as the Baby Grand cathedral. Philco was also in the auto radio business by now, as they had acquired the Automobile Radio Corporation and were offering Transitone auto radios through Philco dealers.

At the seventh annual Radio World's Fair, Philco displayed the "world's largest radio." It was a giant sized replica of a model 96 lowboy and used twenty-two tubes. The set was also shown in other areas of the country. Another giant Philco set, a replica of a model 20 Baby Grand Console, was used at the Omaha Auditorium in Omaha, Nebraska to receive broadcasts of the 1930 World Series.

Late in the year, Philco expanded its Philadelphia plant and built a plant in Toronto, Ontario for the Canadian market. Philco gave its popular model 20 cathedral a facelift; a curly maple front arch was added as well as columns on either side of the front panel.

Philco also promoted its President, Edward Davis, to the newly created position of Chairman of the Board. Philco's General Manager, James M. Skinner, became the company's new President. The two men would supervise Philco through the company's "golden years" - 1930 to 1938.

The sales promotions, tie-ins with Paramount, and displays of giant radios helped stir up interest in Philco, but there is no doubt that it was the model 20 that made Philco number one by the end of the year. Philco's goal of reaching the top in radio sales had been achieved. The question now was, would Philco be able to maintain its lead, since it was a "midget" set that had made them the industry's leader?

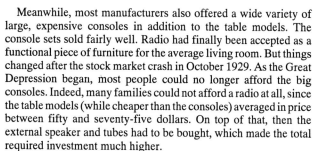

The First Baby Grand

By late 1929, all radio manufacturers' table model sets had become remarkably similar. Many were housed in metal cabinets to cut costs, while a few others were still being made with wood cases. All of them used external speakers, a basic style that had been in use since the early twenties.

A Philco advertising sign with a neon tube around its edge. *Photo by Michael Prosise.*

Meanwhile, most manufacturers also offered a wide variety of large, expensive consoles in addition to the table models. The console sets sold fairly well. Radio had finally been accepted as a functional piece of furniture for the average living room. But things changed after the stock market crash in October 1929. As the Great Depression began, most people could no longer afford the big consoles. Indeed, many families could not afford a radio at all, since the table models (while cheaper than the consoles) averaged in price between fifty and seventy-five dollars. On top of that, then the external speaker and tubes had to be bought, which made the total required investment much higher.

In response to the depression, some small companies (such as Jackson-Bell) began to offer compact table model receivers with built-in speakers. The new radios were quickly dubbed "midget" sets. Today, we call the large table model radios with curved tops "cathedral" radios.

The new midget sets sold from between fifty and sixty dollars, cheaper than most common table model receivers...and they did not require the purchase of an external speaker!

The new midget style was being ignored by most radio manufacturers. They caught the attention of the Philadelphia Storage Battery Company (Philco), however, which saw possibilities of increased sales and perhaps a good shot at the number one spot in the radio industry with the midget radio design.

Philco had fought back from near extinction twice before: in 1906 when they began to make storage batteries and in 1928 when they switched from Socket-Power units to radios. The company was now in third place among radio makers, and was facing another serious challenge to its existence due to depressed business conditions.

As before, Philco met the challenge head on. While they aggressively promoted their new 1930 models that were introduced in June, they prepared to release a midget set of their own. In August, Philco's new Model 20 was introduced to the public. It was a high quality midget set at a competitive price - $49.50, less tubes.

The new receiver, which Philco named the Baby Grand, was a TRF set using seven tubes and an eight inch electrodynamic speaker. Philco advertising claimed the Baby Grand would perform as well as "big" sets. And, true to Philco's word, it did work quite well; better than many other midgets and nearly as well as Philco's own model 77.

Other major radio manufacturers reacted to Philco's entry into the midget set market by rejecting the new trend. Atwater Kent, Crosley and RCA continued to offer the same tired, old fashioned table model sets with separate speakers. Others, such as Majestic and Zenith, chose to continue specializing in consoles.

The end result was that, while most other manufacturers were looking the other way, Philco's Baby Grand was becoming an enormous success. It was just what the public was looking for as the depression settled over America. In a November magazine advertisement, Philco claimed that tens of thousands of Baby Grands were being sold each week. At one point in the fall, Philco's factory had over 140,000 unfilled orders for Baby Grand sets.

Eventually, over 300,000 Baby Grands were made. And the set did what Philco hoped it would do - it made them number one in radio. Other manufacturers would soon realize the need for smaller, cheaper radios.

Note: From this point on, Philco models designed to operate on 115 volts AC, 25-40 cycles, have an "A" suffix after the model number; e.g. Model 20-A, etc.

Model 20
Circuit: Screen Grid TRF
Frequency Coverage: 550-1500 kc
Power: AC; 115 volts, 50-60 cycles
Tubes Used: 7
Controls: 3; Tuning, Off-On, Volume
See Appendix II (4-1) for the tube diagram for Model 20.

Model 20 cathedral. *Photo by Ron Boucher.*

Model 20 consolette. *Photo by Michael Prosise.*

Model 30
Circuit: Screen Grid Plus TRF
Frequency Coverage: 550-1500 kc
Power: Battery operated
Tubes Used: 8
Controls: 4; Tuning, Tone, Off-On, Volume
See Appendix II (4-2) for the tube diagram of Model 30.

Model 30 was available in cabinets identical to Model 96
lowboy and highboy, shown on page 24.

Model 41
Circuit: Screen Grid TRF
Frequency Coverage: 550-1500 kc
Power: 115 volts DC
Tubes Used: 6
Controls: 4; Tuning, Tone, Off-On, Volume
See Appendix II (4-3) for the tube diagram of Model 41.

Model 41 was available in cabinets identical to Model 77
console, shown below, and Model 96 lowboy and highboy, shown on page 24.

Model 77
Circuit: Screen Grid TRF
Frequency Coverage: 550-1500 kc
Power: AC; 115 volts, 50-60 cycles
Tubes Used: 7
Controls: 4; Tuning, Tone, Off-On, Volume
See Appendix II (4-4) for the tube diagram for Model 77.

Model 20 Deluxe cathedral.

The Console. *Courtesy of Michael Prosise.*

Model 96 lowboy.

The Highboy. *Courtesy of Michael Prosise.*

Model 77 lowboy.

Model 77 was also available in a cabinet identical to Model 96 table model, shown on page 24.

Models 96, 296 & Concert Grand
(all used the same chassis)
Circuit: Screen Grid Plus TRF
Frequency Coverage: 550-1500 kc
Power: AC; 115 volts, 50-60 cycles
Tubes Used: 9
Controls: 4; Tuning, Tone, Off-On, Volume
See Appendix II (4-5) for the tube diagram for Model 96, 296 & Concert Grand.

The Concert Grand Radio-Phonograph. *Courtesy of Michael Prosise.*

Model 296 Radio-Phonograph. *Photo by Michael Prosise.*

The Table Model.

Due largely to the huge success of the model 20 cathedral, Philco was now the leading radio manufacturer in the country. So for 1931, what could they do for an encore?

First and foremost, they introduced the cathedral cabinet that has remained a classic up to this day. Designed by Edward L. Combs, the cabinet was first used to house model 21, introduced early in the year as the successor to the highly popular model 20.

But Philco was not content to be thought of as a company that seemed to specialize in "midget" receivers. Late in 1930, RCA began to make its superheterodyne patents available to its licensees. Early in 1931, Philco introduced its first superhet, the 111 Superheterodyne-Plus. It replaced the 96, 296 and the Concert Grand as Philco's new top of the line receiver. It used eleven tubes and was available in a lowboy cabinet (identical to the model 96 lowboy), a new highboy cabinet with double doors, and as the new model 211 radio-phonograph.

The superheterodyne circuit had been invented during World War I by Major Edwin Howard Armstrong, one of radio's greatest inventors, and offered performance far superior to TRF or Neutrodyne receivers. RCA acquired the rights to Armstrong's superheterodyne patents in the early twenties, and for many years refused to license other companies to use the superheterodyne circuit due to its superior performance.

Another new model Philco introduced in early 1931 was the 220 radio-phonograph, using the same chassis as model 20 had. Model 42, a DC powered radio, was also available; it was practically the same as the model 41 it replaced. Models 30 and 77 were still available as well.

Philco's 1931-32 lineup was introduced in June. The new top end models were the 112 lowboy and highboy, and the 212 radio-phonograph. These used virtually the same chassis as the 111 and 211 had, but were housed in new cabinets that were the work of noted industrial designer Norman Bel Geddes.

The new model line included three other new superheterodyne sets, models 70 and 90, and the battery operated model 35. Also available was a new DC powered set, model 46, which had the same basic circuit that had been used in model 20. A lower priced radio-phonograph (model 270) was also available, as well as another Philco "first," the model 370 chairside radio. It was in a cabinet that had also been designed by Norman Bel Geddes.

A significant aspect of Philco's new lineup is that more cabinet styles were available among the various models. This was a departure from Philco's practice of cabinet sharing during the past three years.

Philco had begun research into television in 1928. The company began work on an all-electronic system in 1931. Mechanical TV systems were in use on a very limited basis during this time; having been developed mainly by Charles Francis Jenkins in the U.S. and by John Logie Baird in Britain. However, mechanical television would soon prove to be inferior to all-electronic television.

Mechanical television was transmitted by a fairly complicated method. Light from the "televised scene" passed through one of several holes in a rapidly spinning disk, known as the "scanning disk." This light was picked up by a photocell, which changed the light to an electrical signal. The signal was ultimately transmitted, like a radio signal, and picked up by a scanning disk television receiver. The reciever picked up the signal, which was amplified and used to drive a neon bulb, which shone through another rapidly spinning scanning disk onto a 1 to 2 inch screen. The end result was a very crude picture. Jenkins' televised images were merely silhouettes, while Baird's system would actually show (albeit crudely) subjects.

Of course, electronic television is what we have today. Images were formerly picked up by a special camera tube (now solid state devices are used). The light is changed to an electrical signal, transmitted, and received by the TV receiver. The signal is amplified and sent to a cathode-ray tube or picture tube, which shows the images on its screen.

In the summer of 1931, television inventor and researcher Philo T. Farnsworth joined Philco's TV research team. Farnsworth is best remembered for his invention of the Image Dissector television camera tube. Others were hard at work on electronic television, including Vladimir Zworykin (working for RCA) in the U.S., and Manfred Von Ardenne in Germany.

Meanwhile, Philco had added a new, cheaper radio to its line by October (the model 50 TRF set), a three tube, self-powered, short-wave converter (model 4), and the model 570 grandfather clock radio, which used the same chassis as models 70, 270 and 370.

By now, several radio manufacturers had fallen by the wayside due to the deepening depression. Even Grigsby-Grunow (Majestic) had begun to slide downward. However, the depression seemed to be having the opposite effect on Philco, as they strengthened their hold on the number one spot in 1931 while selling close to a million radios. Philco was also one of only a few companies that made money that year.

Philco's Classic Cathedral Radio Design

For many years now, Philco's model 90 cathedral has been the most popular old radio among collectors and noncollectors alike. In more recent years, many old radios have been reproduced as transistorized AM-FM sets. But it was Philco's model 90 that was the first to be reproduced, as Philco-Ford issued an AM-FM replica of the model 90 in the early 1970s. At that time old radio collecting was not nearly as popular as it is today.

However, Philco's model 90 was not the first set to be offered in this particular cathedral style, which became the most famous of radio designs. To trace the history of this classic set, we have to go back to November 1930, when the first Baby Grand - model 20 - made the Philadelphia Storage Battery Company number one in radio sales. Meanwhile, as model 20 sets continued to sell, work began on its successor. By March 1931, Model 21 was released to the public.

Inside, this new Baby Grand was identical to the model 20, since it also used seven tubes in a TRF circuit. But its cathedral cabinet, designed by Edward L. Combs, was destined to become a classic. Combs was responsible for the design of several Philco cabinets, including the 30, 41, and 96 highboy, the 296 lowboy radio-phonograph, the later models 90X and 112X (Philco's first Inclined Sounding Board consoles), and the 15X console cabinet. Philco filed for a design patent for the new cathedral cabinet on February 28, 1931, not long before the new model 21 went on sale. U.S. design patent number 83,956 was issued on April 21, 1931.

Early model 21 sets used the same knobs as model 20 receivers had used. Later, Philco began to use the rosette knobs on model 21. In late production models, two type 45 audio output tubes were used instead of the type 71A tubes used previously.

In June of 1931, Philco introduced its new 1931-32 line. Model 21 was discontinued, and was replaced by two new sets: Model 70 and 90. The new Baby Grands continued to use Edward Combs' cathedral design. Model 70 was the same size as model 21, and also used seven tubes; model 90 was slightly larger and used nine tubes. The two new Baby Grands performed better than model 21, as both used the superheterodyne circuit. They also had tone controls, which model 21 lacked.

Two other Philco cathedrals, also introduced in June 1931, used the Combs cabinet as well. Model 35, a battery operated 7-tube superheterodyne, was identical to model 70. Model 46 operated on 110 volts direct current (DC), and used six tubes in a TRF circuit. It resembled model 21.

Altogether, five Philco models used the cathedral cabinet designed by Edward L. Combs. It is interesting that, while model 90 is the best known example of this style, it was not the best selling version. More than two and a half times as many model 70 cathedrals (288,620) were made as model 90 Baby Grands (106,050). The reason was simple - price. Model 70 sold for $49.95, while model 90 cost $69.95.

Today, although model 90 is the most sought after example of Combs' cathedral design, models 21 and 70 are also popular among old radio enthusiasts. Collectively, the three are among the most popular of all cathedral radios. Models 35 and 46 are not as popular as the AC operated versions, either because they are scarce, or because they do not operate on alternating current.

Edward Combs probably never dreamed his cathedral design would become the best known and most popular of all cathedral radios.

Model A clock timer, which could be set to turn your radio on and off at preset times. *Photo by Ron Boucher.*

Model 4 Shortwave Converter
Circuit: Superheterodyne
Frequency Coverage: Three bands.
    Band #1: 1.5-3.6 mc
    Band #2: 3.6-8.5 mc
    Band #3: 8.5-19.0 mc
Power: AC; 115 volts, 50-60 cycles
Tubes Used: 3
Controls: 3; Tuning, Off-On, Band Switch
See Appendix II (5-1) for the tube diagram for the Model 4 Shortwave Converter.

Philco's shortwave converter. *Photo by Michael Prosise.*

Model 21
Circuit: Screen Grid TRF
Frequency Coverage: 550-1500 kc
Power: AC; 115 volts, 50-60 cycles
Tubes Used: 7
Controls: 3; Tuning, Off-On, Volume
See Appendix II (5-2) for the tube diagram for Model 21.

Model 35
Circuit: Superheterodyne
Frequency Coverage: 550-1500 kc
Power: Battery operated
Tubes Used: 7
Controls: 4; Tuning, Tone, Off-On, Volume
See Appendix II (5-3) for the tube diagram for the Model 35 cathedral.

Model 35 was available in cabinets identical to Model 70 cathedral and highboy, shown on page 28.

Model 42
Circuit: Screen Grid TRF
Frequency Coverage: 550-1500 kc
Power: 115 volts DC
Tubes Used: 6
Controls: 4; Tuning, Tone, Off-On, Volume
See Appendix II (5-4) for the tube diagram for Model 42.

Model 42 was available in cabinets identical to Model 77 console, shown in Chapter Four on page 24, and Model 96 lowboy and highboy, shown in Chapter Four on page 24.

Model 46
Circuit: Screen Grid TRF
Frequency Coverage: 550-1500 kc
Power: 115 volts DC
Tubes Used: 6, with one ballast
Controls: 3; Tuning, Off-On, Volume
See Appendix II (5-5) for the tube diagram for Model 46.

Model 46 was available in cabinets similar to Model 70 cathedral and highboy, shown on page 28.

Model 50
Circuit: Screen Grid TRF
Frequency Coverage: 550-1500 kc
Power: AC; 115 volts, 50-60 cycles
Tubes Used: 5
Controls: 3; Tuning, Off-On, Volume
See Appendix II (5-6) for the tube diagram for Model 50.

An early version of the model 21 cathedral.

A later version of the Model 21 cathedral. Notice the different knobs on this set. *Photo by Ron Boucher.*

Model 50 cathedral. *Photo by Paul Rosen.*

5-TUBE LOWBOY $49.95

Model 50 lowboy.

Models 70, 270, 370 & 570
(all used the same chassis)
Circuit: Superheterodyne
Frequency Coverage: 550-1500 kc
Power: AC; 115 volts, 50-60 cycles
Tubes Used: 7
Controls: 4; Tuning, Tone, Off-On, Volume.
Model 270 has a separate volume control for the phonograph.
Model 370 has tuning and off-on/volume controls on top of cabinet, with tone control inside.
See Appendix II (5-7) for the tube diagram for Models 70, 270, 370 & 570.

HIGHBOY   Model 70
7-Tube Superheterodyne $65.75
Complete with tubes
Including New Pentode Power Tube

Exquisitely done in figured American Walnut and quilted Maple. Scroll-carved Arch and pin stripe Pilasters. Tone Control. New Electro-Dynamic Speaker. Illuminated Station Recording Dial.
D. C. (Direct Current) Model 46, **$76.75**

Model 70 highboy. *Courtesy of Michael Prosise.*

RADIO-PHONOGRAPH
Model 70 7-Tube Superheterodyne $110
Complete with tubes
Including New Pentode Power Tube

All-Electric. Gracefully designed in figured Black Walnut, and scroll-carved Arch. Tone Control. New Electro-Dynamic Speaker. Illuminated Station Recording Dial.

Model 270 Radio-Phonograph. *Courtesy of Michael Prosise.*

Model 70 cathedral. *Photo by Ron Boucher.*

Model 370 chairside set.

28

Model 570 grandfather clock-radio. *Photo by Michael Prosise.*

A group picture of three Philco cathedrals designed by Edward L. Combs. Left to right: Model 21, Model 70, Model 90. *Photo by Ron Boucher.*

Model 90

Circuit: Superheterodyne
Frequency Coverage: 550-1500 kc
Power: AC; 115 volts, 50-60 cycles
Tubes Used: 9 (see variations below)
Controls: 4; Tuning, Tone, Off-On, Volume

See Appendix II (5-8) for the type 1 tube diagram for Model 90.

See Appendix II (5-9) for the type 2 tube diagram for Model 90.

Model 90 lowboy.

Model 90 cathedral. *Photo by Ron Boucher.*

Closed (left) and open (right) views of the Model 90 highboy.

Models 111 & 211

(both used the same chassis)
Circuit: Superheterodyne-Plus
Frequency Coverage: 550-1500 kc
Power: AC; 115 volts, 50-60 cycles
Tubes Used: 11
Controls: 4; Tuning, Tone, Off-On, Volume; also a Normal-Maximum switch on top of chassis

See Appendix II (5-10) for the tube diagram for Models 111 & 211.

Model 111 lowboy.

Closed (left) and open (right) views of the Model 111 highboy. *Photos by Michael Prosise.*

Model 211 is a radio-phonograph console using the 111 chassis.

Models 112 & 212
(both used the same chassis)
Circuit: Superheterodyne-Plus
Frequency Coverage: 550-1500 kc
Power: AC; 115 volts, 50-60 cycles
Tubes Used: 11 (see variations below)
Controls: 4; Tuning, Tone, Off-On, Volume; also a
Normal-Maximum switch on top of chassis.
See Appendix II (5-11) for the type 1 tube diagram for
Models 112 & 212.

See Appendix II (5-12) for the type 2 tube diagram for
Models 112 & 212.

Model 112 highboy.

Model 220
Circuit: Screen Grid TRF
Frequency Coverage: 550-1500 kc
Power: AC; 115 volts, 50-60 cycles
Tubes Used: 7
Controls: 3; Tuning, Off-On, Volume.
Also has an additional volume control for the phonograph.

See Appendix II (5-13) for the tube diagram for Model
220.

Model 220 is a radio-phonograph console using the same
chassis as Model 20. Model 20 is covered in Chapter
Four on pages 22 and 23.

Model 112 lowboy.

11-TUBE RADIO-PHONOGRAPH   Model 212
11-Tube Superheterodyne-Plus $295.00
Complete with tubes

All-Electric, with Automatic Record Changer. A superb
creation by Norman Bel Geddes. Beautifully executed in
matched Butt Walnut, with details of pin stripe American
Walnut. Instrument Panel with tapestry designed by
Norman Bel Geddes. Sliding doors. Hand-rubbed
finish. 4-Point Tone Control—Automatic Volume Control—Oversize Electro-Dynamic Speaker—Long Distance
Switch—Two 45 Power Tubes—Push-Pull.

◆ ◆ ◆

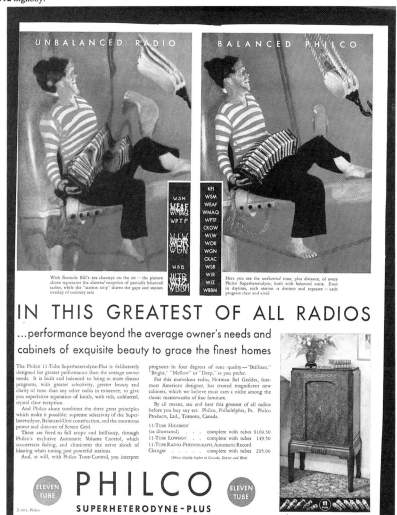

Model 212 Radio-Phonograph. *Courtesy of Michael
Prosise.*

31

The year 1932 would prove to be the worst yet for the radio industry. However, this did not stop Philco from continuing to advertise heavily or from introducing new innovations.

Philco began the new year with more new sets. The five-tube model became a superheterodyne (model 51). A mantle clock with the 51 chassis was also offered, the model 551 Colonial Clock. Two novel consoles were offered, combining a version of the model 4 shortwave converter with either the model 70 or model 90. These were sold as models 470 and 490.

The firm also made some changes in existing models as well. AVC was made a part of model 70 and two type 47 audio output tubes were added to model 90. Both the 70 and 90 now used the new type 35 variable-mu screen grid tubes, which were designed to take advantage of AVC circuits. These types of tubes will amplify weak signals as an ordinary screen grid tube would, with a small AVC voltage applied to them. Under strong signal conditions when more AVC voltage is applied, the variable-mu tube amplifies less, and does so much better than a common screen grid tube such as a type 24.

Philco also began work on high fidelity reception in 1932. The company's first steps toward this goal were models 90X and 112X. These sets had a totally new cabinet design, (which was the work of Edward L. Combs), quite different from other console radio cabinets. The speaker was mounted at an upward angle which was intended to make high frequency notes more audible. At the same time, the large board the speaker was mounted on (which Philco dubbed the "Inclined Sounding Board") helped the low audio frequencies. A screen was placed in back to prevent sound waves from coming out of the back of the cabinet.

Models 4, 35, 46, 212, 270, 370 and 570 were still available in early 1932.

In June, Philco introduced several new models. Most of the new models became the industry's first receivers to use the new 6.3 volt tubes. The new tubes used less filament current than the 2.5 volt tubes they replaced. In addition, many of the new tubes were smaller than the older ones. Meanwhile, most other manufacturers began to use another new line of tubes which were 2.5 volt tubes that still consumed a relatively high amount of filament current. Other radiomakers would soon catch on to the superiority of the new 6.3 volt tubes.

Philco's best receiver this season was model 15. It was very similar to model 112 which it replaced, but used the new 6.3 volt tubes and had a local-distance switch as part of the off-on switch. It was the last Philco model which used the large, heavy chassis which had been used in various versions since 1928. This new set was available in two versions. Model 15X was an inclined sounding board console (yet another Edward Combs design), similar to the previous model 112X. Model 15DX looked a lot like the 15X, but also featured tambour doors which, when closed, hid the control panel.

The new models 91 and 71 were also similar to models 90 and 70 which they replaced (with the new tubes, of course). These were available in a variety of cabinets, including a new cathedral style cabinet. All Philco cathedral models would use this same basic style for the next year and a half. Model 52, which used the same chassis as the earlier model 51, was available in cathedral and lowboy cabinets as well as a new table model design.

Another new Philco feature, used on models 15, 91 and the 23X radio-phonograph, was shadow tuning. This was a miniature tuning meter enclosed in a metal case. A tiny vane functioned similar to the pointer on a conventional meter. Light from a pilot lamp entered the

metal case from the back through a small slot. A translucent plastic screen was mounted at the front of the unit. The light shone on this screen, and the vane made a shadow appear in the middle of the screen. When a station was tuned in, the vane moved, causing the shadow pattern to reduce in width. Philco used shadow meters on many of its models through 1938. They were also used by Atwater Kent and Zenith on some of their sets for a few years.

Philco's first all-wave receiver was now available. Model 43 received the broadcast band and shortwave up to 20 mc, in four bands. The company also offered a battery operated set (model 36) and a new DC powered receiver (model 47). The Norman Bel Geddes chairside was still available, with the new 71 chassis (model 71LZ). Besides the 23X radio-phonograph (which used the 91 chassis), the 22L radio-phonograph was also available. It used the 71 chassis. Models 4, 551 and 570 were also still being offered.

All Philco consoles except models 36 and 52 now had twin speakers. Many radio manufacturers were following this trend. It was short-lived for Philco, however. They had ceased the practice by the next year.

In June 1932, Philco received a license for experimental television station W3XE, one of America's first all-electronic TV stations. Research into television was continuing, by Philco, RCA and others.

By now, Philco had divided into two companies: the Philadelphia Storage Battery Company and the Philco Radio & Television Corporation. Philco did this in an effort to reduce patent royalty payments to RCA. The Philadelphia Storage Battery Company handled most of the manufacturing while the Philco Radio & Television Corporation's purpose was mainly to sell the products.

A model 23X radio-phonograph appeared in Bing Crosby's hotel room in the 1932 Paramount picture, *The Big Broadcast.* Paramount and Philco had advertised together in the spring and fall of 1930. Crosby, the star of the movie, was already a well known popular singer and would later become the star of Philco's own radio broadcast, *Philco Radio Time,* and be featured in later Philco advertising.

By fall, a new receiver had been added to the Philco line. Model 80 used a four-tube reflex circuit in which its second detector also provided intermediate frequency (IF) amplification through a principle known as regeneration. The regenerative circuit had been invented by Major Edwin H. Armstrong, who had also invented the superheterodyne circuit.

Another Philco "first" introduced in the fall was a chairside radio with a separate speaker cabinet (model 14LZX). Two more radio-phonographs were also added to the line (models 24L and 25L), another battery set (model 37) and another radio for DC operation (model 48).

Philco remained number one in 1932, but only sold a little over 600,000 radios during the year.

The company began 1933 by continuing to offer most of the June 1932 line and making more additions as well. New models included the 19 and 89. The only difference in the two is that model 19 has a shadow meter, while model 89 does not. Two more radio-phonographs made their debut (models 26L and 27L), along with two compact table models (the 53C and 54C), and a new four tube cathedral radio with two-band coverage (model 81).

Models 14 and 91 now had two-band coverage also, from the broadcast band through approximately three megacycles. A short time later, models 19 and 89 were changed to two-band coverage as well.

And now let us take a look at what Philco offered during this period.

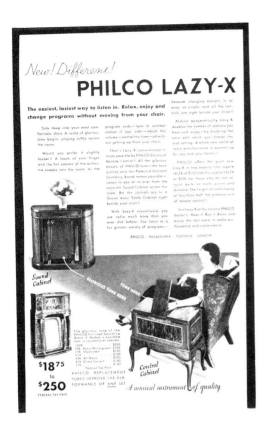

now enjoy console radio sound with the convenience of chairside operation, hence the name "Lazy." All Philco models using the Inclined Sounding Board were identified with an "X" suffix, so that is how the 14LZX came to be called "Lazy X."

Philco continued to offer chairside radios with separate speaker cabinets through 1936.

Extension speaker for use with Philco receivers, from 1932. *Photo by Michael Prosise.*

Note: From this point on, all Philco models use a superheterodyne circuit except where noted.

Models 14, 23 & 91
(all used the same chassis)
Frequency Coverage:
Early production 23 & 91: 550-1500 kc
Later 23, 91 and early 14: 520-1500 kc
Late production 14, 23 & 91: Two bands.
    Band #1: 520-1500 kc
    Band #2: 1.5-3.2 mc
Power: AC; 115 volts, 50-60 cycles
Tubes Used: 9
Controls: 4; Tuning, Tone, Off-On (combined with band switch on late production), Volume
These models use shadow meters.

See Appendix II (6-1) for the tube diagram for Models 14, 23 & 91.

The Philco Lazy X

No, it wasn't a dude ranch for Philco executives; the Lazy X was a new and unique way of providing a radio owner with chairside remote control.

In the late twenties, many radio manufacturers were beginning to offer some type of remote control with their more expensive receivers. Most used some type of control box, which was connected to the radio by a multiwire cable. Eventually, General Motors would offer an add-on remote control for its receivers (model 281) which served three functions. It converted older General Motors radios to remote control operation; it also converted those same GM sets, which used a TRF circuit, to superheterodyne operation; and it had a handy ashtray at the top of it. (Yes, General Motors was one of many manufacturers that were making radios for the home in the early thirties.)

Philco had taken a different direction in allowing operation of a radio from the easy chair by introducing a chairside radio (model 370) in June 1931. The same cabinet, designed by Norman Bel Geddes, housed the new 71LZ which was introduced in June 1932.

In late 1932, Philco introduced its version of "remote control." The new nine-tube model 14LZX featured not only the radio controls, but the entire receiver chassis in a Queen Anne chairside cabinet. The big difference between this set and Philco's earlier chairsides, however, was that the new 14LZX used a separate speaker, connected to the control cabinet by a twenty-five foot cable.

The speaker cabinet looked like a console radio with the top part cut off. The reason for such a large speaker cabinet was to make the Lazy X (Philco's name for the 14LZX) sound like a console radio. The cabinet housed two speakers mounted on Philco's exclusive Inclined Sounding Board.

By early 1933, Philco had introduced another Lazy X model. Model 19LZX used six tubes, and was mounted in cabinets identical to the 14LZX.

The Philco Lazy X models allowed the set owner to place the control cabinet beside his or her favorite chair while the speaker cabinet could be placed across the room. Also, the set owner could

Closed (left) and open (right) views of Model 14LZX
chairside set (separate speaker may be seen on page 33).

Model 91L lowboy. *Courtesy of Michael Prosise.*

Model 23X Radio-Phonograph console. *Courtesy of
Michael Prosise.*

Model 91B cathedral. *Photo by Gary Schieffer.*

Model 91D highboy with doors. *Courtesy of Michael Prosise.*

Model 91X console.

Model 15

Frequency Coverage: 550-1500 kc
Power: AC; 115 volts, 50-60 cycles
Tubes Used: 11
Controls: 4; Tuning, Tone, Off-On/Local-Distance, Volume
These models use shadow meters.
See Appendix II (6-2) for the tube diagram for Model 15.

Model 15X console. *Photo by Wayne King.*

Model 15DX console. *Photo by Spencer Doggett.*

Models 19, 26, 27 & 89

(all used the same chassis)
Frequency Coverage:
Early production: 540-1500 kc
Late production: Two bands.
    Band #1: 540-1500 kc
    Band #2: 1.5-3.2 mc
Power: AC; 115 volts, 50-60 cycles
Tubes Used: 6
Controls:
Early production: 3; Tuning, Tone, Off-On/Volume
Late production: 4; Tuning, Tone, Off-On/Band Switch, Volume
Models 19 and 26 use shadow meters.
See Appendix II (6-3) for the tube diagram for Models 19, 26, 27 & 89.

The more common version of the AM-only model 89B cathedral.

Model 19B cathedral. *Photo by Bob Schafbuch.*

Model 89B cathedral (early version).

Model 89L lowboy.

Model 19L is identical to Model 89L lowboy shown above.

Model 19LZ is in a chairside cabinet.

Model 19LZX is in cabinets identical to Model 14LZX chairside set with separate speaker, shown on pages 33 & 34.

Model 26L is a radio-phonograph console using the 19 chassis.

Model 27L is a radio-phonograph console using the 89 chassis.

Models 22 & 71
(both used the same chassis)
Frequency Coverage:
    Early production: 550-1500 kc
    Late production: 520-1500 kc
Power: AC; 115 volts, 50-60 cycles
Tubes Used: 7
Controls: 4; Tuning, Tone, Off-On, Volume
Model 71LZ has tuning and off-on/volume controls on top of cabinet, with tone control inside.
See Appendix II (6-4) for the tube diagram for Models 22 & 71.

Model 71L lowboy. *Courtesy of Michael Prosise.*   Model 71H highboy. *Courtesy of Michael Prosise.*

Model 22L Radio-Phonograph. *Courtesy of Michael Prosise.*

Model 71LZ chairside.

Model 71D is in a cabinet identical to Model 91D highboy with doors, shown on page 35.

Models 24, 51, 52 & 551
(all used the same chassis)
Frequency Coverage:
    All except late production 52: 550-1500 kc
    Late production 52: 540-1500 kc
Power: AC; 115 volts, 50-60 cycles
Tubes Used: 5
Controls: 3; Tuning, Off-On, Volume
See Appendix II (6-5) for the tube diagram of Models 24, 51, 52 & 551.

Model 71B cathedral. *Photo by John Okolowicz.*

Model 51B cathedral. *Photo by Jerry McKinney.*

Model 52B cathedral.

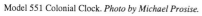

Model 551 Colonial Clock. *Photo by Michael Prosise.*

Model 52C table model. *Courtesy of Michael Prosise.*

Model 24L is a radio-phonograph console using the 52 chassis.

Models 51L & 52L are in cabinets identical to Model 37L lowboy, shown on page 39.

Models 25 & 43
(both used the same chassis)
Frequency Coverage: Four bands.
    Band #1: 550-1400 kc
    Band #2: 1.4-3.4 mc
    Band #3: 3.3-9.0 mc
    Band #4: 8.5-20.0 mc
Power: AC; 115 volts, 50-60 cycles
Tubes Used: 8
Controls: 4; Tuning, Off-On/Volume, Tone, Band Switch
See Appendix II (6-6) for the tube diagram for Models 25 & 43.

Model 43B cathedral. *Courtesy of Michael Prosise.*

Model 25L is a radio-phonograph console using the 43 chassis.

Model 43H is in a cabinet identical to Model 71H highboy, shown on page 37.

Model 43X is in a cabinet similar to Model 91X console, shown on page 35.

Model 36
Frequency Coverage: 550-1500 kc
Power: Battery operated
Tubes Used: 7
Controls: 4; Tuning, Tone, Off-On, Volume
See Appendix II (6-7) for the tube diagram for Model 36.

Model 36B is in a cabinet identical to Model 71B cathedral, shown on page 37.

Model 36L is in a cabinet identical to Model 71L lowboy, shown on page 37.

Model 36D is in a cabinet identical to Model 91D highboy with doors, shown on page 35.

Model 37
Frequency Coverage: 540-1500 kc
Power: Battery operated
Tubes Used: 5, with one ballast
Controls: 3; Tuning, Volume, Off-On
See Appendix II (6-8) for the tube diagram for Model 37.

Model 37L lowboy.

Model 37C is in a cabinet identical to Model 52C table model, shown on page 38.

Model 47
Frequency Coverage:
    Early production: 550-1500 kc
    Late production: 520-1500 kc
Power: 115 volts DC
Tubes Used: 8
Controls: 4; Tuning, Tone, Off-On, Volume
See Appendix II (6-9) for the tube diagram for Model 47.

Model 47B is in a cabinet similar to Model 91B cathedral, shown on page 34.

Model 47H is in a cabinet similar to Model 71H highboy, shown on page 37.

Model 47D is in a cabinet similar to Model 91D highboy with doors, shown on page 35.

Model 47X is in a cabinet similar to Model 91X console, shown on page 35.

Model 48
Frequency Coverage: 540-1500 kc
Power: 115 volts DC
Tubes Used: 4, with one ballast
Controls: 3; Tuning, Volume, Off-On
See Appendix II (6-10) for the tube diagram for Model 48.

Model 48C is in a cabinet identical to Model 52C table model, shown on page 38.

Model 48L is in a cabinet identical to Model 37L lowboy, shown on page 39.

Models 53 & 54
Frequency Coverage:
    Model 53: 540-1500 kc
    Model 54: Two bands.
    Band #1: 540-1500 kc
    Band #2: 1.5-3.2 mc
Power: 115 volts AC/DC
Tubes Used: see below
Controls: 2; Off-On/Volume, Tuning
Model 54 also has a band switch in back of set
See Appendix II (6-11) for the tube diagram for Model 53.
See Appendix II (6-12) for the tube diagram for Model 54.

Model 54C table model. *Photo by Paul Rosen.*

Model 53C is in a cabinet similar to Model 54C table model shown above.

Models 70, 270, 370 & 570
(all used the same chassis)
Frequency Coverage: 550-1500 kc
Power: AC; 115 volts, 50-60 cycles
Tubes Used: 7
Controls: 4; Tuning, Tone, Off-On, Volume
Model 370 has tuning and off-on/volume controls on top
of cabinet, with tone control inside
See Appendix II (6-13) for the tube diagram for Models
70, 270, 370 & 570.

For pictures of these models, refer to Models 70, 270, 370
& 570 in Chapter Five on pages 28 & 29.

Model 80
Frequency Coverage: 540-1500 kc
Power: AC; 115 volts, 50-60 cycles
Tubes Used: 4
Controls: 2; Tuning, Off-On/Volume
See Appendix II (6-14) for the tube diagram for Model
80.

Model 80 Colonial Clock. *Photo by Ron Boucher.*

Model 80B cathedral.

Model 80C table model.

Model 80 portable receiver. *Photo by Doug Houston.*

Model 81
Frequency Coverage: Two bands.
    Band #1: 540-1500 kc
    Band #2: 1.5-2.8 mc
Power: AC; 115 volts, 50-60 cycles
Tubes Used: 4
Controls: 4; Tuning, Band Switch, Off-On, Volume
Later production models have three controls as the off-on switch was combined with the volume control
See Appendix II (6-15) for the tube diagram for Model 81.

Model 81B is in a cabinet similar to Model 80B cathedral, shown on page 40.

Model 90
Frequency Coverage: 550-1500 kc
Power: AC; 115 volts, 50-60 cycles
Tubes Used: 9
Controls: 4; Tuning, Tone, Off-On, Volume
See Appendix II (6-16) for the type 3 tube diagram for Model 90.

Model 90X console.

Model 90 was also available in a cathedral cabinet, shown in Chapter Five on page 29.

Model 112X
Frequency Coverage: 550-1500 kc
Power: AC; 115 volts, 50-60 cycles
Tubes Used: 11
Controls: 4; Tuning, Tone, Off-On, Volume; also a Normal-Maximum switch on top of chassis
See Appendix II (6-17) for the tube diagram for Model 112X.

Model 112X console.

Model 470
Frequency Coverage:
    Broadcast unit: 550-1500 kc
    Shortwave unit: Three bands.
    Band #1: 1.5-3.6 mc
    Band #2: 3.6-8.5 mc
    Band #3: 8.5-19.0 mc
Power: AC; 115 volts, 50-60 cycles
Tubes Used: 9 (see variations below)
Controls: 7.
Top (broadcast) unit: Tuning, Tone, Off-On, Volume
Bottom (shortwave) unit: Tuning, Off-On, Band Switch
See Appendix II (6-18) for the tube diagram for the Model 470 Shortwave unit.

See Appendix II (6-19) for the tube diagram for the Model 470 Broadcast unit (early).

See Appendix II (6-20) for the tube diagram for the Model 470 Broadcast unit (late).

Model 470 console. *Photo by John Sedlacek.*

Model 490
Frequency Coverage:
  Broadcast unit: 550-1500 kc
  Shortwave unit: Three bands.
  Band #1: 1.5-3.6 mc
  Band #2: 3.6-8.5 mc
  Band #3: 8.5-19.0 mc
Power: AC; 115 volts, 50-60 cycles
Tubes Used: 11 (see variations below)
Controls: 7.
Top (broadcast) unit: Tuning, Tone, Off-On, Volume
Bottom (shortwave) unit: Tuning, Off-On, Band Switch
See Appendix II (6-21) for the tube diagram for the Model 490 Shortwave unit.

See Appendix II (6-22) for the tube diagram for the Model 490 Broadcast unit (early).

See Appendix II (6-23) for the tube diagram for the Model 490 Broadcast unit (late).

Model 490 console. *Photo by John Miller.*

Philco's new 1933-34 line was introduced in June 1933. These were mostly all new sets; the only "leftovers" from the previous season were models 19, 43, 47, 54 and 89.

The new top of the line model was also one of the best radios Philco ever made. The all-wave model 16 used eleven tubes and covered, in five bands, standard broadcast and shortwave from 520 kc to 23 mc. The two console models and the chairside version used a large twelve inch speaker with a big, heavy field coil which gave these models very good fidelity. A new "Super Class A" audio output circuit with an output of fifteen watts (ten in the 16B cathedral) also gave the radio better fidelity. This new circuit, with a type 42 pentode tube connected as a triode and driving two type 42 tubes (also connected as triodes) in push-pull, was also used in the new models 14, 17 and 18 as well as the new 500, 501 and 503 radio-phonographs.

Many other manufacturers had adopted class B audio output circuits the previous year. The class B circuit gave a higher power output, but also produced more distortion than class A circuits. Philco rejected class B audio for AC sets, although they did adopt it for use in battery operated receivers. Once again, other manufacturers would soon realize Philco was right and drop class B audio in AC sets.

Since a large portion of the listening public were now interested in shortwave reception, Model 16 sold well. It was an excellent receiver. Of course, the fact that it sounded quite good did not hurt either. Philco was getting closer to its goal of high fidelity sound.

An interesting set is model 17. It also used eleven tubes, but only covered the broadcast band and shortwave from 1.5 to 4 mc. Even the smaller eight-tube model 43 had greater shortwave capability than the 17, as it was another all-wave model covering 550 kc to 20 mc in four bands. The limited shortwave capability in model 17 was also available in cheaper models such as the 14, 18 and 60. Model 17 would not return the next season.

Another new feature shared by models 16 and 17 was a squelch control, which Philco called Quiet Automatic Volume Control, or QAVC. It could be switched on or off by a toggle switch on the side of the cabinet. When on, it would reduce or even eliminate noise between stations. A control on the back of the chassis adjusted the level at which the QAVC became effective. For some reason, Philco only offered QAVC during the 1933-34 season, and for a few months on model 16 at the beginning of the 1935 season. Today, most communications sets such as CB radios have squelch controls.

Other new Philco models included the battery operated model 38, which was sold in cabinets identical to the new AC powered model 60, and a new look-alike to the five-tube model 54C, the four-tube model 57C. Philco also offered four new radio-phonographs; model 500X (with the 16 chassis), 501X (with the 16 chassis and an automatic record changer), 503L (with the 18 chassis), and the 505L (with the 60 chassis).

In the fall, model 43 was replaced with another new all-wave set; the six-tube model 44. It covered standard broadcast and shortwave from 520 kc to 23 mc in four bands, and its chassis was also available in the 504L radio-phonograph.

Philco sales bounced back in 1933. The company sold nearly a million radios, about as many as they had sold in 1931.

All Philco cathedrals still shared the same styling that had been introduced in June 1932. This began to change at mid-season. Early in 1934, Philco introduced new cathedral cabinets for models 38 and 60. A similar cabinet was used for the new model 84B.

Philco Baby Grand design entered a new phase at mid-season, as the company introduced its first tombstone style radios. Model 16B was put into a large, Art Deco tombstone cabinet; the set now used a ten inch speaker instead of the eight inch unit used in the 16B cathedral model. The Baby Grand 16 had become a sort of miniature console, designed to sit on a table!

Model 60 was now also available in a Moderne tombstone cabinet as model 60MB. The Moderne styling was also used in a new console cabinet for model 14, known as the 14MX.

The year 1934 would be Philco's best yet, as they would sell one and a quarter million radios and capture thirty percent of the American radio market.

The Philco Model 16

In the early thirties, radio listeners were becoming aware of signals from faraway lands on the shortwave frequencies. Nearly a decade earlier, set owners often stayed up late at night in order to see if their receiver would give them coast-to-coast reception. Now, it was becoming possible for the listener to tune in signals from all over the world.

Philco had marketed a shortwave converter in late 1931, and had introduced its first all-wave receiver (model 43) in June 1932.

In 1934, according to the Radio Manufacturers' Association, an all-wave receiver was a set which would pick up all the frequencies from 540 kc to 18 mc. This guideline was adopted by the RMA after some manufacturers falsely claimed that some of their sets were all-wave sets, when they might have had only one shortwave band with limited capability.

In June 1933, Philco's new 1933-34 lineup was led by another all-wave set. However, model 16 was more than just an all-wave radio. It used eleven tubes and included such features as Philco's new Super Class A audio circuit, with fifteen watt output in console and chairside models, and ten watt output in the 16B. Other features of the 16 included two stages of IF amplification and a squelch control, which Philco called Quiet Automatic Volume Control (QAVC). Model 16 also had five bands, covering frequencies from 520 kc to 23 mc.

An eleven tube chassis in a cathedral cabinet? Yes, and the 16B was one of the largest radios designed to sit on a table that was ever made. It used an eight inch speaker, while model 16 consoles and the chairside version with separate speaker cabinet used twelve inch speakers.

At mid-season, Philco put the 16 chassis in an even larger Art Deco tombstone cabinet, along with a larger ten inch speaker.

While Philco was extolling the virtues of model 16's high quality tone, customers were buying them for their shortwave reception. Model 16's AM and shortwave performance is quite good due to the powerful eleven tube circuit. The shadow meter helps to tune in stations correctly, while the QAVC reduces noise when tuning.

Excellent shortwave performance, pleasing tone, and good looks made model 16 an outstanding set, whether it was the lowboy 16L console, Inclined Sounding Board 16X, chairside 16RX with separate speaker, 16B cathedral or 16B tombstone. It is most certainly one of the best shortwave radios ever made.

Model 14L lowboy. *Photo by Wayne King.*

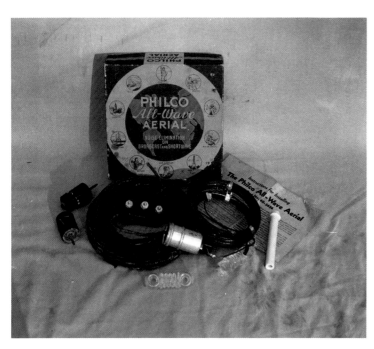

Philco All-Wave Aerial kit from the mid thirties. *Photo by Michael Prosise.*

Model 14X console.

14X — 18X
Slightly smaller versions of the Inclined Sounding Board model in a hand-rubbed walnut cabinet that is a masterpiece of grace and beauty. The art of the past, as revealed in outstanding examples in the Metropolitan Museum, has contributed to the development of a design which is at once classic and yet a distinctively modern treatment of a musical instrument. Figured Striped and Butt Walnut is combined with delicate mouldings and marquetry to produce in this cabinet an effect never before achieved in cabinet design.

Model 14
Frequency Coverage: Two bands.
    Band #1: 520-1500 kc
    Band #2: 1.5-4.0 mc
Power: AC; 115 volts, 50-60 cycles
Tubes Used: 9
Controls: 4; Tuning, Off-On/Volume, Band Switch, Tone
These models use shadow meters.
See Appendix II (7-1) for the tube diagram for Model 14.

Model 14MX console. *Photo by Michael Prosise.*

Model 14B is in a cabinet similar to Model 16B cathedral, shown below.

Model 14RX is in cabinets identical to Model 16RX chairside set with separate speaker, shown on page 46.

Models 16, 500 & 501

(all used the same chassis)

Frequency Coverage: Five bands.

    Band #1: 520-1500 kc

    Band #2: 1.5-4.0 mc

    Band #3: 3.2-6.0 mc

    Band #4: 5.8-12.0 mc

    Band #5: 11.0-23.0 mc

Power: AC; 115 volts, 50-60 cycles

Tubes Used: 11

Controls: 4; Tuning, Off-On/Volume, Band Switch, Tone; also a QAVC on-off switch on side of cabinet and a QAVC control on back of chassis

These models use shadow meters and feature Quiet Automatic Volume Control (QAVC)

See Appendix II (7-2) for the tube diagram for Models 16, 500 & 501.

Model 16L lowboy.

Model 16B cathedral.

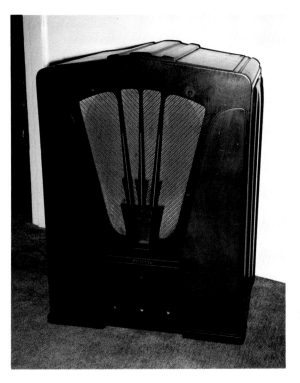

Model 16X console. *Photo by John Okolowicz.*

Model 16B tombstone. *Photo by Jerry McKinney.*

45

Model 17D highboy with doors.

17D
18D

A six legged Highboy cabinet of unusual beauty with diamond matched Oriental wood pilasters. Doors of beautifully matched Butt Walnut on the outside and highly figured Oriental wood on the inside. Quilted Maple panels above and below doors.

Model 16RX chairside set (left) with separate speaker (right).

Models 500X & 501X are radio-phonograph consoles using the 16 chassis. Model 501X features an automatic record changer.

16X
17X

Patented Inclined Sounding Board models with exquisite cabinet of hand rubbed walnut with curved panels of figured Striped Walnut on the base. At the top curved panels of Butt Walnut. Inlays of Black and Satin wood enhance the beauty of this super example of furniture craftsmanship.

Model 17X console.

Model 17

Frequency Coverage: Two bands.
 Band #1: 520-1500 kc
 Band #2: 1.5-4.0 mc
Power: AC; 115 volts, 50-60 cycles
Tubes Used: 11
Controls: 4; Tuning, Off-On/Volume, Band Switch, Tone; also a QAVC on-off switch on side of cabinet and a QAVC control on back of chassis
These models use shadow meters and feature Quiet Automatic Volume Control (QAVC)
See Appendix II (7-3) for the tube diagram for Model 17.

Model 17L lowboy (late version). *Photo by Spencer Doggett.*

Model 17RX chairside set. Separate speaker used with 17RX is identical with that used with model 16RX, shown on page 46. *Photo by Spencer Doggett.*

Model 17B is in a cabinet similar to Model 16B cathedral, shown on page 45.

Model 17L (early version) is in a cabinet similar to Model 14L lowboy, shown on page 44.

Models 18 & 503
(both used the same chassis)
Frequency Coverage: Two bands.
    Band #1: 520-1500 kc
    Band #2: 1.5-4.0 mc
Power: AC; 115 volts, 50-60 cycles
Tubes Used: 8
Controls: 4; Tuning, Off-On/Volume, Band Switch, Tone
These models use shadow meters.
See Appendix II (7-4) for the tube diagram for Models 18 & 503.

Model 18L lowboy.

18H

A new and attractive Highboy of hand-rubbed walnut trimmed with Oriental wood and inlays of Satinwood. One of the most popular conventional cabinets in the whole PHILCO line.

Model 18H highboy (early version).

Model 18H highboy (late version).

Model 19B cathedral.

Model 18X console. *Photo by Spencer Doggett.*

Model 18B is in a cabinet similar to Model 16B cathedral, shown on page 45.

Model 18D is in a cabinet similar to Model 17D highboy with doors, shown on page 46.

Model 18RX is in cabinets identical to Model 16RX chairside set with separate speaker, shown on page 46.

Model 503L is a radio-phonograph console using the 18 chassis.

Models 19 & 89
(both used the same chassis)
Frequency Coverage: Two bands.
    Band #1: 540-1500 kc
    Band #2: 1.5-3.2 mc
Power: AC; 115 volts, 50-60 cycles
Tubes Used: 6
Controls: 4; Tuning, Tone, Off-On/Band Switch, Volume
Model 19 uses a shadow meter.
See Appendix II (7-5) for the tube diagram for Models 19 & 89.

Model 19H highboy.

Model 19TX chairside set (separate speaker not shown).
Photo by Michael Prosise.

Note the slight difference in these two model 89B cathedrals. *Photo at left by Dennis Osborne; at right by Gary Schieffer.*

Model 89L is in a cabinet similar to the early 1933 Model 89L lowboy, shown in Chapter Six on page 36.

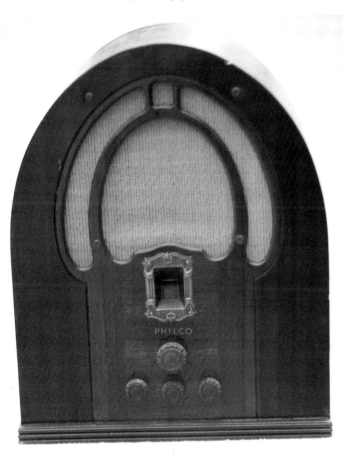

Models 38 & 38-A
Frequency Coverage: Two bands.
    Band #1: 520-1500 kc
    Band #2: 1.5-2.4 mc
Power: Battery operated
Tubes Used: 5
Controls: 4; Tuning, Volume, Band Switch, Off-On
Model 38-A uses an additional ballast tube.
See Appendix II (7-6) for the tube diagram for Models 38
& 38-A.

Model 38B cathedral (early version).

Model 38B cathedral (late version).

Model 38L is in a cabinet similar to Model 18L lowboy,
shown on page 47.

Model 43
Frequency Coverage: Four bands.
    Band #1: 550-1500 kc
    Band #2: 1.4-3.4 mc
    Band #3: 3.3-9.0 mc
    Band #4: 8.5-20.0 mc
Power: AC; 115 volts, 50-60 cycles
Tubes Used: 8
Controls: 4; Tuning, Off-On/Volume, Tone, Band Switch
See Appendix II (7-7) for the tube diagram for Model 43.

Model 43B cathedral.

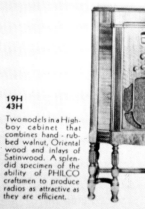

Model 43H highboy.

Models 44 & 504
(both used the same chassis)
Frequency Coverage: Four bands.
    Band #1: 520-1500 kc
    Band #2: 1.5-4.0 mc
    Band #3: 4.0-11.0 mc
    Band #4: 11.0-23.0 mc
Power: AC; 115 volts, 50-60 cycles
Tubes Used: 6
Controls: 4; Tuning, Off-On/Volume, Band Switch, Tone
See Appendix II (7-8) for the tube diagram for Models 44
& 504.

Model 47
Frequency Coverage: 520-1500 kc
Power: 115 volts DC
Tubes Used: 8
Controls: 4; Tuning, Tone, Off-On, Volume
See Appendix II (7-9) for the tube diagram for Model 47.

Model 47B is in a cabinet similar to Model 43B cathedral, shown on page 50.

Model 47D is in a cabinet similar to Model 17D highboy with doors, shown on page 46.

Model 47H is in a cabinet similar to Model 44H highboy, shown at lower left.

Model 47X is in a cabinet similar to Model 18X console, shown on page 48.

Models 57 & 58
(both used very similar chasses)
Frequency Coverage: 540-1700 kc
Power: AC; 115 volts, 50-60 cycles
Tubes Used: 4
Controls: 2; Off-On/Volume, Tuning
See Appendix (7-10) II for the tube diagram for Models 57 & 58.

Model 44B cathedral. *Photo by David Kendall.*

Model 57C table model. *Photo by Doug Houston.*

Model 58C is in a cabinet similar to Model 57C table model, shown above.

Models 60 & 505
(both used the same chassis)
Frequency Coverage: Two bands.
   Band #1: 530-1500 kc
   Band #2: 1.5-4.0 mc
Power: AC; 115 volts, 50-60 cycles
Tubes Used: 5
Controls: 4; Tuning, Off-On/Volume, Band Switch, Tone
See Appendix II (7-11) for the tube diagram for Models 60 & 505.

Model 44H highboy.

Model 504L is a radio-phonograph console using the 44 chassis.

Model 60B cathedral (first version). *Photo by Ron Boucher.*

Model 60MB tombstone. *Photo by Ron Boucher.*

Model 60L is in a cabinet similar to Model 18L lowboy, shown on page 47.

Model 84

Frequency Coverage: 540-1740 kc
Power: AC; 115 volts, 50-60 cycles
Tubes Used: 4
Controls: 2; Tuning, Off-On/Volume
See Appendix II (7-12) for the tube diagram for Model 84.

Model 84B cathedral. *Photo by Paul Rosen.*

Model 60B cathedral (second version).

Model 505L Radio-Phonograph.

Philco's 1935 model lineup, introduced in June 1934, was larger than ever. It included many all-new models as well as new cabinets for some older models.

The flagship model this year was the ten-tube model 200X. Although this set received only the broadcast band, it was the first real high fidelity radio any manufacturer had ever offered. The new Philco 200X had an audio output of fifteen watts using the Super Class A circuit, and an IF expander which could be adjusted by a "Fidelity-Selectivity" control on the front panel. For the highest fidelity, the control was set to allow a wide IF bandwidth. It could also be adjusted for sharp selectivity which allowed sharper tuning but lowered the fidelity.

Model 16 was still available for 1935. It received new cabinets except for the 16B tombstone which continued from the previous season. The new model 18 had been slightly modified; it now received the broadcast band only. An almost identical set, the new model 118, received the broadcast band and also shortwave from 4.2 to 12 mc. Last year's model 44 received a shadowmeter and new cabinets, and was now called model 144. Other Philco models from the previous season, unchanged except for some cabinets, were models 38, 54, 60, 84 and 89.

Other all new Philco models for the 1935 season included models 28, 29, 45, 59 and 66; the battery operated model 34; the DC powered model 49; and model 32 for thirty-two volt DC farm systems.

Six radio-phonographs were offered by Philco for 1935. These included models 500X and 501X, both using the 16 chassis; model 503L, using the 18 chassis; model 505L, using the 60 chassis; model 506L, using the 144 chassis; and model 507L, using the 118 chassis. Models 500X and 501X were similar, with the exception being the automatic record changer used in model 501X. Model 500X used a manual electric phonograph.

In October, Philco introduced a slightly modified model 16. It now had four bands while providing nearly the same frequency coverage (540 kc to 22.5 mc), and no longer had QAVC.

By now it was common practice for Philco to make changes and additions to its model lineup at mid-season. In January, a new high fidelity model was offered. Model 201X was similar to the 200X, with the addition of a 4.2 to 12 mc shortwave band. Philco also offered a "limited edition" radio-phonograph, model 509X, which used the 201 chassis and an automatic record changer. According to Philco advertising of the period, only five hundred 509X models were to be made.

Philco was beginning to slowly phase out the cathedral style cabinets as models 32, 34, 49, 118 and 144 received new tombstone cabinets.

Also at mid-season, Philco introduced a new battery operated receiver (model 39) and introduced new console cabinets for several other models.

Not only was Philco's lineup growing, but the company's sales were increasing as well as they sold one and a half million radios in 1935.

Courtesy of Michael Prosise.

Philco Achieves High Fidelity

To anyone accustomed to listening to an AM station on a modern receiver, it may seem impossible to think that AM receivers were, at one time, capable of producing much greater fidelity than today's cheap sets.

Since 1932, Philco had been trying to achieve high fidelity sound by using such innovations as their exclusive Inclined Sounding Board and Super Class A audio. However, the real answer to high fidelity AM was to broaden the IF bandwidth of the receiver. This would allow a higher audio frequency response, at the expense of reduced selectivity. Therefore, to allow fidelity and selectivity, it had to be possible to make the IF bandwidth variable.

So, in June 1934, Philco introduced the first high fidelity receiver. Model 200X used ten tubes and included variable IF bandwidth, by means of a "Fidelity-Selectivity" control on the front panel. It only received the AM band, but it could reproduce audio frequencies from 50 to 7,500 cycles, which was considered at the time to be "high fidelity." (Today, we consider high fidelity to be 20 to 20,000 cycles.)

While Philco took most of the credit for the breakthrough in high fidelity reproduction, they received help in developing their variable IF bandwidth circuit from the Hazeltine Laboratories.

Midway through the 1935 season, Philco added two more high fidelity receivers to its lineup. Model 201X received AM as well as one shortwave band, which covered 4.2 to 12 mc. Model 509X used the 201 chassis in a console radio-phonograph combination.

The new high fidelity receivers made possible a more lifelike sound. Philco's high fidelity sets seem to work best on a strong AM signal. Although I have not heard a 200X or 201X in operation, I own a model 37-116X which was made in 1937, and is also a high fidelity receiver with variable IF bandwidth. On strong AM signals, the 37-116X produces a sound that rivals FM!

If you know someone who owns one of Philco's high fidelity receivers, ask him or her to let you listen to it. After you have heard what AM sets are capable of doing, you will know better when you hear someone say it is not possible for AM stations to sound very good!

Models 16, 500 & 501
(these used the same chassis)
Frequency Coverage: Four bands.
    Band #1: 550-1500 kc
    Band #2: 1.5-4.1 mc
    Band #3: 4.1-10.0 mc
    Band #4: 9.8-22.5 mc
Power: AC; 115 volts, 50-60 cycles
Tubes Used: 11
Controls: 4; Tuning, Off-On/Volume, Band Switch, Tone; also a bass compensation switch on right side of cabinet
These models use shadow meters.
See Appendix II (8-1) for the tube diagram for Models 16, 500 & 501.

Model 16B tombstone (early version). *Photo by Doug Houston.*

A Philco advertising clock from the mid thirties. *Photo by Michael Prosise.*

Model 16B tombstone (late version). *Photo by Michael Prosise.*

Model 16X console. *Photo by Doug Houston.*

16L—American and Foreign All-Wave
34L—Battery—American and Foreign All-Wave

Model 16L lowboy.

Model 16RX chairside set (left) with separate speaker (right).

Model 500X Radio-Phonograph.

Model 501X is in a cabinet identical to Model 500X shown above, and features an automatic record changer.

Please note that from June until October, Model 16 still used the 1933-34 5-band chassis in the new 1935 cabinets (early version only on 16B tombstone).

Models 18, 118, 503 & 507
(these used very similar chasses)
Frequency Coverage:
  Models 18 & 503: 530-1720 kc
  Models 118 & 507: Two bands.
  Band #1: 540-1720 kc
  Band #2: 4.2-12.0 mc
Power: AC; 115 volts, 50-60 cycles
Tubes Used: 8
Controls: 4.
Models 18 & 503: Tuning, Volume, Off-On, Tone Models 118 & 507: Tuning, Off-On/Volume, Band Switch, Tone
These models use shadow meters.
See Appendix II (8-2) for the tube diagram for Models 18, 118, 503 & 507.

Model 118B tombstone. *Photo by Jerry McKinney.*

Model 118B cathedral. *Photo by Ron Boucher.*

Model 118H highboy. *Photo by Edmund DeCann.*

PHILCO 118D
American and Foreign Broadcast Receiver

Model 118D highboy with doors (early version).

Model 3118X console (late version). This is a Canadian model, identical to the U.S. model 118X. *Photo by Paul Rosen.*

118D—American and Foreign
28D—AC-DC—American and Foreign
49D—DC—American and Foreign

Model 118D highboy with doors (late version).

Model 118MX console.

Model 118RX chairside set (left) with separate speaker cabinet (right).

Model 507L Radio-Phonograph console. *Photo by Michael Prosise.*

Model 18B was offered in cathedral and tombstone cabinets identical to Model 118B cathedral and tombstone.

Model 18H is in a cabinet identical to Model 118H highboy.

Model 118X (early version) is in a cabinet identical to Model 18X console, shown in Chapter Seven on page 48.

Model 18MX is in a cabinet identical to Model 118MX console.

Model 503L is in a cabinet identical to Model 507L radio-phonograph.

Model 28
Frequency Coverage: Two bands.
    Band #1: 540-1720 kc
    Band #2: 4.2-13.0 mc
Power: 115 volts AC/DC
Tubes Used: 6
Controls: 4; Tuning, Off-On/Volume, Tone, Band Switch
  See Appendix II (8-3) for the tube diagram for Model 28.

Model 28C is in a cabinet identical to Model 45C table model, shown on page 60.

Model 28L is in a cabinet similar to Model 89L lowboy, shown on page 64.

Model 28D is in a cabinet similar to Model 118D highboy with doors (late version), shown on page 57.

Model 28F is in a cabinet identical to Model 45F console, shown on page 60.

Model 28CSX is in a cabinet similar to model 635CSX chairside set, shown in Chapter Nine on page 74.

Models 29 & 45
(both used very similar chasses)
Frequency Coverage: Two bands.
    Band #1: 540-1720 kc
    Band #2: 4.2-13.0 mc
Power: AC; 115 volts, 50-60 cycles
Tubes Used: 6
Controls: 4; Tuning, Off-On/Volume, Tone, Band Switch
Model 29 uses a shadow meter.
See Appendix II (8-4) for the tube diagram for Models 29 & 45.

PHILCO 99X

Model 29X console (early version).

Model 29X console (late version).

Separate speaker cabinet used with both versions of Model 29TX.

Early version (left) and late version (right) of Model 29TX chairside set. *Photo at left by Jerry McKinney; at right by Michael Prosise.*

Model 45C table model. *Photo by Ron Boucher.*

Model 45L lowboy.

Model 45F console. *Photo by Michael Prosise.*

Model 29CSX is in a cabinet similar to Model 635CSX chairside set, shown in Chapter Nine on page 74.

Model 32
Frequency Coverage: Two bands.
    Band #1: 540-1500 kc
    Band #2: 1.5-3.2 mc
Power: 32 volts DC
Tubes Used: 6 (see variations below)
Controls: 4; Tuning, Tone, Off-On/Band Switch, Volume
See Appendix II (8-5) for the tube diagram for Model 32.

Model 32B cathedral is in a cabinet similar to Model 118B cathedral, shown on page 56.

Model 32B tombstone is in a cabinet similar to Model 144B tombstone, shown on page 64.

Model 32L is in a cabinet similar to Model 89L lowboy, shown on page 64.

Models 34 & 34-A
(these used the same chassis)
Frequency Coverage: Four bands.
    Band #1: 520-1500 kc
    Band #2: 1.5-4.0 mc
    Band #3: 4.0-11.0 mc
    Band #4: 11.0-23.0 mc
Power: Battery operated
Tubes Used: 7
Controls: 4; Tuning, Off-On/Volume, Band Switch, Tone
Model 34-A uses an additional ballast tube.
See Appendix II (8-6) for the tube diagram for Models 34 & 34-A.

Model 34B cathedral is in a cabinet similar to Model 118B cathedral, shown on page 56.

Model 34B tombstone is in a cabinet similar to Model 144B tombstone, shown on page 64.

Model 34L is in a cabinet similar to Model 16L lowboy, shown on page 55.

Models 38 & 38-A

(these used the same chassis)
Frequency Coverage: Two bands.
  Band #1: 520-1500 kc
  Band #2: 1.5-2.4 mc
Power: Battery operated
Tubes Used: 5
Controls: 4; Tuning, Volume, Band Switch, Off-On
Model 38-A uses an additional ballast tube.
See Appendix II (8-7) for the tube diagram for Models 38 & 38-A.

Model 38B is in a cabinet identical to Model 60B cathedral, shown on page 62.

Model 38L is in a cabinet similar to Model 89L lowboy, shown on page 64.

Models 39 & 39-A

(these used the same chassis)
Frequency Coverage: Two bands.
  Band #1: 550-1720 kc
  Band #2: 5.5-16.0 mc
Power: Battery operated
Tubes Used: 6
Controls: 4; Tuning, Off-On/Volume, Band Switch, Tone
Model 39-A uses an additional ballast tube.
See Appendix II (8-8) for the tube diagram for Models 39 & 39-A.

Model 39B cathedral.

Model 39F is in a cabinet similar to Model 45F console, shown on page 60.

Model 49

Frequency Coverage: Two bands.
  Band #1: 530-1720 kc
  Band #2: 4.2-12.0 mc
Power: 115 volts DC
Tubes Used: 7
Controls: 4; Tuning, Off-On/Volume, Band Switch, Tone
These models use shadow meters.
See Appendix II (8-9) for the tube diagram for Model 49.

Model 49B cathedral is in a cabinet identical to Model 118B cathedral, shown on page 56.

Model 49B tombstone is in a cabinet identical to Model 118B tombstone, shown on page 56.

Model 49D (early version) is in a cabinet identical to Model 118D highboy with doors (early version), shown on page 57.

Model 49D (late version) is in a cabinet identical to Model 118D highboy with doors (late version), shown on page 57.

Model 49H is in a cabinet identical to Model 118H highboy, shown on page 56.

Model 49X (early version) is in a cabinet identical to Model 18X console, shown in Chapter Seven on page 48.

Model 49X (late version) is in a cabinet identical to Model 3118X console (late version), shown on page 57.

Models 54 & 59

Frequency Coverage:
  Model 54: Two bands.
  Band #1: 540-1500 kc
  Band #2: 1.5-3.2 mc
  Model 59: 540-1720 kc
  Power:
  Model 54: 115 volts AC/DC
  Model 59: AC; 115 volts, 50-60 cycles
Tubes Used: see below
Controls: 2; Off-On/Volume, Tuning
Model 54 also has a band switch in back of set
See Appendix II (8-10) for the tube diagram for Model 54.
See Appendix II (8-11) for the tube diagram for Model 59.

Model 54C table model (late version). *Photo by Doug Houston.*

Model 54S table model. *Photo by Michael Prosise.*

Model 54C (early version) is shown in Chapter Six on page 39.

Model 59C and 59S are in cabinets identical to Model 54C and 54S.

Models 60 & 505

(both used the same chassis)
Frequency Coverage: Two bands.
　　Band #1: 530-1500 kc
　　Band #2: 1.5-4.0 mc
Power: AC; 115 volts, 50-60 cycles
Tubes Used: 5
Controls: 4; Tuning, Off-On/Volume, Band Switch, Tone
See Appendix II (8-12) for the tube diagram for Models
60 & 505.

Model 60B cathedral. *Photo by Doug Houston.*

Model 60L lowboy.

Model 505L radio-phonograph is shown in Chapter
Seven on page 52.

Model 66

Frequency Coverage: Two bands.
　　Band #1: 540-1720 kc
　　Band #2: 5.5-16.0 mc
Power: AC; 115 volts, 50-60 cycles
Tubes Used: 5
Controls: 4; Tuning, Off-On/Volume, Band Switch, Tone
See Appendix II (8-13) for the tube diagram for Model
66.

Model 66B tombstone. *Photo by Doug Houston.*

Model 66B cathedral. *Photo by Doug Houston.*

Model 84B cathedral (late version). *Photo by Wayne King.*

Model 89

Frequency Coverage: Two bands.
    Band #1: 540-1500 kc
    Band #2: 1.5-3.2 mc
Power: AC; 115 volts, 50-60 cycles
Tubes Used: 6 (see variations below)
Controls: 4; Tuning, Tone, Off-On/Band Switch, Volume (on later models, the off-on switch is combined with the volume control)
See Appendix II (8-15) for the tube diagram for Model 89.

Model 66S tombstone. *Photo by Gary Schieffer.*

Model 66L is in a cabinet similar to Model 89L lowboy, shown on page 64.

Model 84

Frequency Coverage: 540-1740 kc
Power: AC; 115 volts, 50-60 cycles
Tubes Used: 4
Controls: 2; Tuning, Off-On/Volume
See Appendix II (8-14) for the tube diagram for Model 84.

Model 89B cathedral. *Photo by Ron Boucher.*

Model 84B cathedral (early version). *Photo by Doug Houston.*

Model 89L lowboy.

Model 97

Frequency Coverage: Three bands.
  Band #1: 550-1750 kc
  Band #2: 1.75-5.75 mc
  Band #3: 5.75-18.0 mc
Power: AC; 115 volts, 50-60 cycles
Tubes Used: 7
Controls: 4; Tuning, Off-On/Volume, Band Switch, Tone
See Appendix II (8-16) for the tube diagram for Model 97.

Models 144 & 506
(both used the same chassis)
Frequency Coverage: Four bands.
  Band #1: 520-1500 kc
  Band #2: 1.5-4.0 mc
  Band #3: 4.0-11.0 mc
  Band #4: 11.0-23.0 mc
Power: AC; 115 volts, 50-60 cycles
Tubes Used: 6
Controls: 4; Tuning, Off-On/Volume, Band Switch, Tone
These models use shadow meters.
See Appendix II (8-17) for the tube diagram for Models 144 & 506.

144B —American and
      Foreign All-Wave
32B —32 Volt— American
34B —Battery— American
      and Foreign All-Wave

Model 144B tombstone.

Model 144B cathedral. *Photo by Wayne King.*

Model 97X console. *Photo by Spencer Doggett.*

Model 97B was offered in a tombstone cabinet.

64

Model 200

Frequency Coverage: 540-1720 kc
Power: AC; 115 volts, 50-60 cycles
Tubes Used: 10
Controls: 4; Tuning, Off-On/Volume, Bass, Fidelity/Selectivity
This model uses a shadow meter and features variable IF bandwidth
See Appendix II (8-18) for the tube diagram for Model 200.

PHILCO 144H

Model 144H highboy.

PHILCO High-Fidelity 200X

Model 200X console.

Models 201 & 509

(both used the same chassis)
Frequency Coverage: Two bands.
    Band #1: 540-1720 kc
    Band #2: 4.2-12.0 mc
Power: AC; 115 volts, 50-60 cycles
Tubes Used: 10
Controls: 4; Tuning, Off-On/Volume, Band Switch, Fidelity/Selectivity; also a bass compensation switch on side of cabinet
These models use shadow meters and feature variable IF bandwidth
See Appendix II (8-19) for the tube diagram for Models 201 & 509.

Model 144X console (early version).

Model 144X console (late version) is in a cabinet similar to Model 3118X console (late version), shown on page 57.

Model 506L is in a cabinet similar to Model 507L radio-phonograph console, shown on page 58.

Model 201X console.

Model 509X is a radio-phonograph console using the 201 chassis.

In June 1935, the radio industry introduced their new 1936 models. The sets offered by most manufacturers featured a new development: metal shell, octal based tubes. However, Philco rejected the new tubes in favor of what they called the "time-tested, trouble-free" glass tubes. Another major manufacturer, Zenith, also chose to continue using glass tubes.

Philco's new 1936 models may not have had any new metal tubes, but the company did continue to introduce new innovations. Their new top end receiver was the largest Philco had produced to date. The fifteen-tube model 680X was another high fidelity receiver with variable IF bandwidth; four-band coverage including long wave, standard broadcast, and two shortwave bands; and the Super Class A audio system, with fifteen watt power output. It not only used a large speaker, but featured another Philco "first" - acoustic clarifiers. These units were similar to conventional speakers, but with no voice coil, field coil or magnet. The audio waves in the sound chamber caused their cones to vibrate, thus assisting the main speaker. The same type of units are used in some of today's high end stereo speaker systems, but they are now known as passive radiators.

Model 680X also used the new 6A3 "high fidelity" audio output tubes (which were glass, of course), a separate bass amplifier with variable bass control, an amplifier for the shadow meter, and an automatic volume control amplifier.

The new receiver also shared some other new features with the rest of the new 1936 Philco models, such as an Automatic Aerial Selector designed for use with the new Philco All-Wave Aerial. The lead-in of the new Philco antenna was a two-wire transmission line that was designed to connect to two new screw terminals on the back of each new Philco set. As the band switch was changed on the receiver, it was also supposed to tune the antenna for best efficiency.

Another new feature was the Precision Radio Dial, with station settings spaced farther apart for easier tuning.

The new model 116 was a slight modification of the 1935 model 16, with the addition of a fifth band that covered long wave frequencies from 150 to 390 kc. The console version of this set also had acoustic clarifiers and the new 6A3 tubes, as did the 116PX radio-phonograph.

Philco offered innovative cabinet styling on several models for the 1936 season. Models 650H and 660L were designed to look more like furniture than radios. In addition, the dial and controls of these sets were hidden under the top lid of the cabinets. Models 650MX, 650RX and 680X also had hidden dials and controls.

Some of Philco's new 1936 consoles, including models 116X, 650MX, 650PX and 660X, were designed by Benjamin S. Nash, who also designed the later version of the 680X.

More conventional cabinets were available for other new models, including the 610, 611, 620, 623, 624, 630, 640, 641, 642, 643, 650 and 660. Model 624 was Philco's first home radio to operate from a six-volt storage battery through a vibrator power supply, such as used on automobile radios.

Philco also offered model 642, another receiver for use on thirty-two volt DC farm supplies.

Models 32, 38, 54, 59, 60, 84 and 89 returned from the previous season. Model 32 had new tombstone and console cabinets for 1936, while models 38, 60 and 89 featured new console cabinets.

At mid-season, models 38B, 60B, 84B and 116B received new cabinets, as did the 680X. Several new models were introduced, which were slightly modified versions of the sets they replaced (models 625, 635, 645, 655 and 665). Other new models at mid-season included models 600, 602, 604 and 651.

Philco sold 1,900,000 sets in 1936, as the company continued to grow and lead the radio industry.

PHILCO 116X

Select Your Radio from 43 New 1936 Philcos
Liberal Trade-in Allowance — EASY TERMS

Glass or Metal Tubes?

Philco's new 1936 models did not have any startling technological breakthroughs, with the possible exception of Acoustic Clarifiers. But then, on the other hand, neither did anyone else.

The only new and unusual item available when the new radio models were introduced in June 1935 was the metal-shell, octal base tube. Many manufacturers adopted the new metal tubes right away, but not Philco!

The company did not see a need to switch to new tubes when glass tubes had been doing a good job. Indeed, the metal tube craze would eventually prove to be frivolous.

When we take a close look at glass and metal tubes, we find that both have their advantages and disadvantages. Metal tubes did not require external metal shields as did some glass tubes, as the metal shell of the tube doubled as a shield. Metal tubes were smaller than the taper-top glass tubes then available, allowing somewhat more compact receiver design.

On the other hand, metal tubes could be broken almost as easily as glass tubes, as they also used glass inside the metal shell. When a glass tube does not light up, it can usually be noticed right away. But you cannot tell by looking at a metal tube whether it is lit or not. Also, when a tube develops gas, it often gives off a blue or purple glow inside the glass. Again, you cannot see inside a metal tube, so a high quality tube tester with a gas test must be used to tell positively if a metal tube is gassy or not. (When a tube develops gas, it often adversely affects its performance.)

Philco never did use metal tubes in its radios. Within a few years, most other radio manufacturers weren't using them either.

Philco "Teleflash" from the mid thirties. Not actually a radio; the cabinet contains an audio amplifier and a speaker. It was wired to telephone lines and may have been used for news reports or horse racing play-by-play. *Photo by Michael Prosise.*

Model 32
Frequency Coverage: Two bands.
    Band #1: 540-1500 kc
    Band #2: 1.5-3.2 mc
Power: 32 volts DC
Tubes Used: 6
Controls: 4; Tuning, Tone, Band Switch, Off-On/Volume
See Appendix II (9-1) for the tube diagram for Model 32.

Model 32F console.

Model 32B is in a cabinet similar to Model 655B tombstone, shown on page 76.

Models 38 & 38-A
(these used the same chassis)
Frequency Coverage: Two bands.
    Band #1: 520-1500 kc
    Band #2: 1.5-2.4 mc
Power: Battery operated
Tubes Used: 5
Controls: 4; Tuning, Volume, Band Switch, Off-On
Model 38-A uses an additional ballast tube.
See Appendix II (9-2) for the tube diagram for Models 38 & 38-A.

Model 38B cathedral (late version). *Photo by Wayne King.*

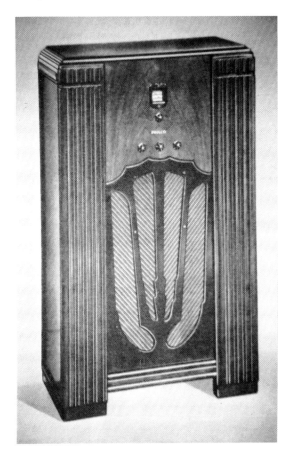

Model 38F console.

Model 38B (early version) is in a cabinet identical to Model 60B cathedral (early version), shown on page 68.

Model 60
Frequency Coverage: Two bands.
    Band #1: 530-1500 kc
    Band #2: 1.5-4.0 mc
Power: AC; 115 volts, 50-60 cycles
Tubes Used: 5
Controls: 4; Tuning, Off-On/Volume, Band Switch, Tone
See Appendix II (9-3) for the tube diagram for Model 60.

Model 60F console.

Model 89
Frequency Coverage: Two bands.
    Band #1: 540-1500 kc
    Band #2: 1.5-3.2 mc
Power: AC; 115 volts, 50-60 cycles
Tubes Used: 6
Controls: 4; Tuning, Tone, Band Switch, Off-On/Volume
See Appendix II (9-4) for the tube diagram for Model 89.

Model 60B cathedral (early version).

Model 60B cathedral (late version).

Model 89B cathedral.

Model 89F console.

Model 116

Frequency Coverage: Five bands.
    Band #1: 150-390 kc
    Band #2: 540-1500 kc
    Band #3: 1.5-4.1 mc
    Band #4: 4.1-10.0 mc
    Band #5: 9.7-22.5 mc
Power: AC; 115 volts, 50-60 cycles
Tubes Used: 11
Controls: 4; Tuning, Off-On/Volume, Band Switch, Tone
These models use shadow meters. Models 116X and
116PX also feature Acoustic Clarifiers.
See Appendix II (9-5) for the tube diagram for Model
116.

Model 116B tombstone (late version). *Photo by Doug Houston.*

Model 116X console.

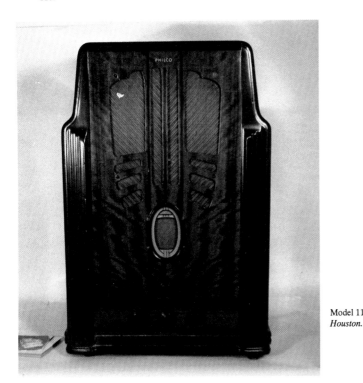

Model 116B tombstone (early version). *Photo by Doug Houston.*

Model 116PX Radio-Phonograph.

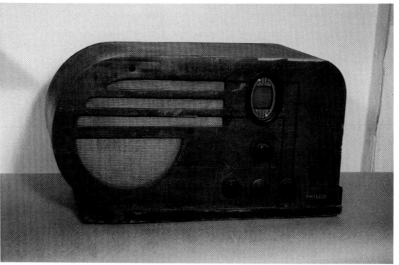

Model 610T table model. *Photo by Doug Houston.*

Model 600 & 602
Frequency Coverage: 530-1800 kc
Power:
    Model 600: AC; 115 volts, 50-60 cycles
    Model 602: 115 volts AC/DC
Tubes Used: see below
Controls: 2; Tuning, Off-On/Volume
See Appendix II (9-6) for the tube diagram for Model 600.
See Appendix II (9-7) for the tube diagram for Model 602.

Models 600C & 602C were available in identical table model cabinets.

Model 604
Frequency Coverage: Two bands.
    Band #1: 530-1750 kc
    Band #2: 6.0-18.0 mc
Power: 115 volts AC/DC
Tubes Used: 5
Controls: 3; Band Switch, Tuning, Off-On/Volume
See Appendix II (9-8) for the tube diagram for Model 604.

Model 604C is in a cabinet similar to Model 37-604C table model, shown in Chapter Ten on page 86.

Model 610
Frequency Coverage: Three bands.
    Band #1: 530-1720 kc
    Band #2: 2.3-2.5 mc
    Band #3: 5.7-18.0 mc
Power: AC; 115 volts, 50-60 cycles
Tubes Used: 5
Controls: 4; Tuning, Off-On/Volume, Band Switch, Tone
See Appendix II (9-9) for the tube diagram for Model 610.

Model 610F console.

Model 610B tombstone.

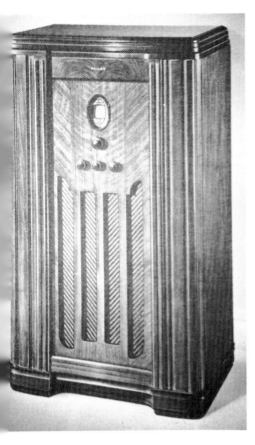

Model 610PF Radio-Phonograph.

Model 610 Radiobar. *Photo by Gary Schieffer.*

Model 611

Frequency Coverage: Three bands.
    Band #1: 530-1720 kc
    Band #2: 2.3-2.5 mc
    Band #3: 5.7-18.0 mc
Power: 115 volts AC/DC
Tubes Used: 5
Controls: 4; Tuning, Off-On/Volume, Band Switch, Tone
See Appendix II (9-10) for the tube diagram for Model 611.

Model 611B tombstone.

Model 611F console.

Models 620 & 625

(both used very similar chasses)
Frequency Coverage: Three bands.
    Band #1: 540-1720 kc
    Band #2: 1.75-5.8 mc
    Band #3: 5.7-18.0 mc
Power: AC; 115 volts, 50-60 cycles
Tubes Used: 6
Controls: 4; Tuning, Off-On/Volume, Band Switch, Tone
See Appendix II (9-11) for the tube diagram for Models 620 & 625.

Model 620B tombstone. *Photo by Michael Prosise.*

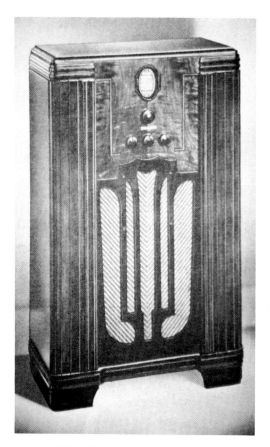

Model 620F console.

(these used the same chassis)
Frequency Coverage: Three bands.
    Band #1: 530-1720 kc
    Band #2: 2.3-2.5 mc
    Band #3: 5.7-18.0 mc
Power: Battery operated
Tubes Used: 6
Controls: 4; Tuning, Off-On/Volume, Band Switch, Tone
Model 623-A uses an additional ballast tube.
See Appendix II (9-12) for the tube diagram for Models
623 & 623-A.

Model 623B tombstone.

Model 625J console.

Model 625B is in a cabinet identical to Model 620B
tombstone.

Model 623F console.

Model 624

Frequency Coverage: Three bands.
    Band #1: 530-1720 kc
    Band #2: 2.3-2.5 mc
    Band #3: 5.7-18.0 mc
Power: 6-volt battery operated
Tubes Used: 6
Controls: 4; Tuning, Off-On/Volume, Band Switch, Tone
See Appendix II (9-13) for the tube diagram for Model 624.

Model 630X console.

Model 624F console.

Model 624B is in a cabinet identical to Model 620B tombstone shown on page 71.

Models 630 & 635

(both used very similar chasses)
Frequency Coverage: Three bands.
    Band #1: 540-1720 kc
    Band #2: 1.75-5.8 mc
    Band #3: 5.7-18.0 mc
Power: AC; 115 volts, 50-60 cycles
Tubes Used: 6
Controls: 4; Tuning, Off-On/Volume, Band Switch, Tone
These models use shadow meters.
See Appendix II (9-14) for the tube diagram for Models 630 & 635.

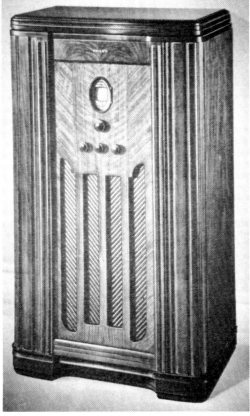

Model 630B tombstone.

Model 630PF Radio-Phonograph.

Model 635J console. *Photo by Doug Houston.*

Models 640 & 645
(both used similar chasses)
Frequency Coverage:
    Model 640: Four bands.
    Band #1: 145-390 kc
    Band #2: 540-1720 kc
    Band #3: 2.2-2.6 mc
    Band #4: 5.8-18.0 mc
    Model 645: Three bands.
    Band #1: 540-1750 kc
    Band #2: 1.75-5.8 mc
    Band #3: 5.75-18.0 mc
Power: AC; 115 volts, 50-60 cycles
Tubes Used: 7
Controls: 4; Tuning, Tone, Band Switch, Off-On/Volume
These models use shadow meters.
See Appendix II (9-15) for the tube diagram for Models 640 & 645.

Model 640B tombstone.

Model 635CSX chairside set. *Photo by Doug Houston.*

Model 635B is in a cabinet identical to Model 630B tombstone.
Model 630CSX is in a cabinet identical to Model 635CSX chairside set.

Model 640X console. *Photo by Doug Houston.*

Model 645 was offered in the same cabinets as Model 640.

Model 641

Frequency Coverage: Three bands.
    Band #1: 530-1720 kc
    Band #2: 2.2-2.6 mc
    Band #3: 5.8-18.0 mc
Power: 115 volts DC
Tubes Used: 7
Controls: 4; Tuning, Tone, Band Switch, Off-On/Volume
These models use shadow meters.
See Appendix II (9-16) for the tube diagram for Model 641.

Model 641B is in a cabinet similar to Model 655B tombstone, shown on page 76.

Model 641X is in a cabinet similar to Model 655X console, shown on page 76.

Model 642

Frequency Coverage: Three bands.
    Band #1: 540-1750 kc
    Band #2: 1.75-5.8 mc
    Band #3: 5.7-18.0 mc
Power: 32 volts DC
Tubes Used: 7
Controls: 4; Tuning, Off-On/Volume, Band Switch, Tone
See Appendix II (9-17) for the tube diagram for Model 642.

Model 642F console.

Model 642B is in a cabinet similar to Model 620B tombstone, shown on page 71.

Models 643 & 643-A

(these used the same chassis)
Frequency Coverage: Four bands.
    Band #1: 150-390 kc
    Band #2: 540-1750 kc
    Band #3: 1.75-5.8 mc
    Band #4: 5.8-18.0 mc
Power: Battery operated
Tubes Used: 7
Controls: 4; Tuning, Tone, Band Switch, Off-On/Volume
Model 643-A uses an additional ballast tube.
See Appendix II (9-18) for the tube diagram for Models 643 & 643-A.

Model 643X console.

Model 643B is in a cabinet similar to Model 655B tombstone, shown on page 76.

Models 650 & 655

(both used similar chasses)
Frequency Coverage:
    Model 650: Four bands.
    Band #1: 145-390 kc
    Band #2: 540-1720 kc
    Band #3: 2.2-2.6 mc
    Band #4: 5.8-18.0 mc
    Model 655: Three bands.
    Band #1: 540-1750 kc
    Band #2: 1.75-5.8 mc
    Band #3: 5.75-18.0 mc
Power: AC; 115 volts, 50-60 cycles
Tubes Used: 8
Controls: 4; Tuning, Tone, Band Switch, Off-On/Volume
These models use shadow meters.
See Appendix II (9-19) for the tube diagram for Models 650 & 655.

Model 655B tombstone. *Photo by Paul Rosen.*

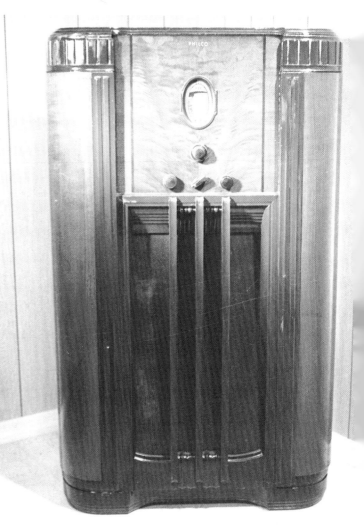

Model 655X console. *Photo by Bob Schafbuch.*

Model 650MX console.

Model 650RX chairside set with separate speaker.

Model 650H console. *Photo by Doug Houston.*

Models 650 & 655 were offered in identical cabinets.

Model 651
Frequency Coverage: Three bands.
    Band #1: 540-1750 kc
    Band #2: 1.75-5.8 mc
    Band #3: 5.75-18.0 mc
Power: 115 volts AC/DC
Tubes Used: 8
Controls: Tuning, Tone, Band Switch, Off-On/Volume
These models use shadow meters.
See Appendix II (9-20) for the tube diagram for Model 651.

Model 651B tombstone. *Photo by John Okolowicz.*

Model 651 was also offered in a console cabinet.

Model 650PX Radio-Phonograph.

Models 660 & 665
(both used very similar chasses)
Frequency Coverage: Four bands.
    Band #1: 150-390 kc
    Band #2: 540-1750 kc
    Band #3: 1.75-5.8 mc
    Band #4: 5.7-18.0 mc
Power: AC; 115 volts, 50-60 cycles
Tubes Used: 10
Controls: 4; Tuning, Off-On/Volume, Band Switch, Tone
These models use shadow meters.
See Appendix II (9-21) for the tube diagram for Models
660 & 665.

Model 660X console.

Model 660L console.

Models 660 & 665 were offered in identical cabinets.

Model 680
Frequency Coverage: Four bands.
    Band #1: 150-400 kc
    Band #2: 550-1700 kc
    Band #3: 2.3-7.0 mc
    Band #4: 7.2-22.0 mc
Power: AC; 115 volts, 50-60 cycles
Tubes Used: 15
Controls: 6; Tuning, Bass, Off-On, Band Switch, Volume, Fidelity/Selectivity
This model uses a shadow meter and features variable IF bandwidth and Acoustic Clarifiers.
See Appendix II (9-22) for the tube diagram for Model 680.

Model 680X console (early version). *Photo by Michael Prosise.*

Model 680X (late version) is in a cabinet which is somewhat similar to model 680X (early version). However, the radio dial and controls are exposed instead of hidden, and are at the upper front of the cabinet.

A new and radically different type of chassis design was adopted for the majority of Philco's new 1937 line, which was introduced in June 1936. Philco called the new design "Unit Construction" and it consisted of a subchassis containing the tuning condenser, band switch, coils, tube(s) and related components, mounted with rubber cushions in the middle of the main chassis, which had the IF and second detector stages on one side, and the audio output and power supply on the other side.

Philco also changed its model numbering system to include a prefix before the model number. The prefix was the last two digits of the model year, so all the new models carried the prefix "37-" this season.

Most of the new Philcos had round escutcheons with a curved window in the middle, through which the dial could be seen; a design that continued to set Philco apart from the rest of the industry.

But despite all these changes, Philco's sales would actually drop in 1937 to just over a million and a half sets. The company remained in the number one position, however.

Philco's three best models were available with a new feature, Automatic Tuning. To tune in a favorite AM station, all that was required was to press in a lever where the station's call letters were located on the set's escutcheon and turn the lever until it clicked into place. The radio was then tuned to the station. To keep the station tuned correctly, Philco introduced automatic frequency control, which they named "Magnetic Tuning."

Philco seemed to be competing with custom set builder E.H. Scott with its new flagship model, the 37-690X. It used twenty tubes, a separate audio output and power supply chassis, a large "Cathedral High Fidelity" speaker, comparable to a modern woofer, and two high frequency speakers, which would now be called tweeters. These were the kind of features found in Scott's expensive receivers.

The 37-690X also had variable IF bandwidth for high fidelity reproduction, Acoustic Clarifiers, variable bass control, Automatic Tuning, and Magnetic Tuning. Its five bands covered frequencies between 530 kc and 18 mc. The set was housed in a large cabinet with double doors.

Model 37-116 had many of the 37-690's features. It had fifteen tubes, variable IF bandwidth, Acoustic Clarifiers, Magnetic Tuning, and the same five-band coverage. It was available with or without Automatic Tuning.

Model 37-675 was the next step down. While not advertised as a high fidelity receiver, it did produce a very good sound due largely to the separate bass and treble controls. It used twelve tubes and was also available with or without Automatic Tuning.

Philco offered its largest lineup to date for 1937. A wide selection of tombstone, console, and table models were available. A few cathedral models were also available this season (models 37-33B, 37-34B, 37-60B, 37-61B, 37-84B and 37-89B). However, Philco advertised no radio-phonograph models for the new season.

For those who wanted to listen to long wave as well as standard broadcast and shortwave, Philco offered three models (the 37-2620, 37-2650 and 37-2670) that had this capability. And if you wanted to hear what was going on between 28 and 42 mc, you would choose the 37-665, which covered these frequencies as well as standard broadcast and some shortwave. All of these models were available in tombstone and console cabinets.

Four of Philco's 1937 models (the 37-84B, 37-600C, 37-602C and 37-604C) used basically the same chassis as the 1936 models 84B, 600C, 602C and 604C, respectively. However, these new sets (along with all other new 1937 Philcos) used octal based glass tubes.

At mid-season, models 37-33B and 37-84B received new cathedral cabinets, while models 37-34B, 37-60B, 37-61B and 37-89B were now offered in tombstone cabinets. Five new models were also introduced. Models 37-9X, 37-10X and 37-11X were consoles with Automatic Tuning and Magnetic Tuning. A new table model was also offered (the 37-62C).

Yet another new model, the 37-93B, is very significant. It was

Philco's last cathedral style radio. It was the end of an era.

The tombstone style cabinet's days were numbered as well. A new era was just around the corner.

Automatic and Magnetic Tuning

Radio manufacturers were beginning to realize that set owners might want to have a convenient means of tuning in favorite standard broadcast stations, without using the tuning knob to select stations manually.

For some reason, the obvious solution of providing pushbuttons to preset favorite stations was overlooked by some manufacturers, as more elaborate mechanical systems were adopted initially. For example, in Grunow Radio's Teledial system, a favorite station was selected by placing a finger in the proper place on the mechanism and then rotating it around the circumference of the dial to the correct position.

Philco introduced its mechanical Automatic Tuning system in its three most expensive 1937 models. It vaguely resembled a telephone dial, as did Grunow's Teledial. Up to nineteen favorite AM stations could be preset in the Philco Automatic Tuning sets.

The mechanism was positioned below the dial, with the tuning knobs in the middle of it. Call letter tabs were placed in the proper slots of the Automatic Tuning dial. To select a favorite station, a lever was rotated to the proper position, pushed in and twirled to the bottom of the mechanism, where it clicked into place. Philco also provided a means to mute the receiver while the Automatic Tuning dial was being operated.

Philco's models 37-675 and 37-116 were initially offered with or without Automatic Tuning. However, all 37-690 sets came with the Automatic Tuning system.

In conjunction with Automatic Tuning, Philco also offered automatic frequency control (AFC), which was called Magnetic Tuning. This was necessary since a station tuned in with the Automatic Tuning mechanism could not always be tuned in precisely, due to inaccuracies in the mechanical system. Adding AFC assured that a station tuned in automatically would be tuned correctly, when the Magnetic Tuning switch was turned on.

Philco's Magnetic Tuning could be used on shortwave as well as the standard broadcast band, as long as the signal was strong. Once the station was tuned in (manually or with the Automatic Tuning mechanism), and the Magnetic Tuning switch turned on, the station would not drift. Unfortunately, Philco's AFC did not work on weak signals.

In the middle of the season, three new Philco consoles (models 37-9, 37-10 and 37-11) were added to the lineup. These three also featured Automatic Tuning and Magnetic Tuning.

Philco reached its peak of electronic design with these models. While all six models were very good receivers, the 37-116 and 37-690 (along with the following season's 38-116 and 38-690), with their high fidelity reproduction, variable IF bandwidth, and acoustic clarifiers in addition to Automatic Tuning and Magnetic Tuning, are definitely among the best radios ever made.

Model 37-9
Frequency Coverage: Three bands.
    Band #1: 530-1720 kc
    Band #2: 2.3-7.4 mc
    Band #3: 7.35-22.0 mc
Power: AC; 115 volts, 50-60 cycles
Tubes Used: 9
Controls: 5; Tuning, Off-On/Tone, Band Switch, Magnetic Tuning, Volume
This model features Automatic Tuning and Magnetic Tuning (AFC)
See Appendix II (10-1) for the tube diagram for Model 37-9.

Model 37-9X console.

Model 37-10
Frequency Coverage: Three bands.
    Band #1: 530-1720 kc
    Band #2: 2.3-7.4 mc
    Band #3: 7.35-22.0 mc
Power: AC; 115 volts, 50-60 cycles
Tubes Used: 9
Controls: 5; Tuning, Off-on/Tone, Band Switch, Magnetic Tuning, Volume
This model features Automatic Tuning and Magnetic Tuning (AFC)
See Appendix II (10-2) for the tube diagram for Model 37-10.

A "Christmas" Philco cardboard sign from the mid thirties.

Model 37-10X console.

Model 37-11
Frequency Coverage: Three bands.
    Band #1: 530-1720 kc
    Band #2: 2.3-7.4 mc
    Band #3: 7.35-22.0 mc
Power: AC; 115 volts, 50-60 cycles
Tubes Used: 10
Controls: 5; Tuning, Off-On/Tone, Band Switch, Magnetic Tuning, Volume
This model features Automatic Tuning and Magnetic Tuning (AFC)
See Appendix II (10-3) for the tube diagram for Model 37-11.

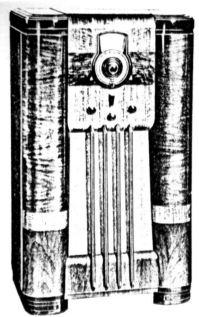

PHILCO 11X

American and Foreign . . Automatic Tuning of favorite stations . .
Magnetic Tuning on Standard Broadcasts . . 3 Tuning Ranges with
Glowing Beam Tuning Range Indicator . . Push-Pull Audio System . .
10 Philco High-Efficiency Tubes . . Philco Color Dial with 50%
greater separation on foreign stations . . 18 Tuned Circuits with the
Philco High-Efficiency Aerial . . Philco Foreign Tuning System.

Model 37-11X console.

Model 37-33
Frequency Coverage: 530-1720 kc
Power: Battery operated
Tubes Used: 5
Controls: 2; Tuning, Off-On/Volume
See Appendix II (10-4) for the tube diagram for Model 37-33.

Model 37-333B cathedral (early version). This is a Canadian model, identical to the U.S. 37-33B. *Photo by Paul Rosen.*

Model 37-33F console.

Model 37-33B (late version) is in a cabinet identical to Model 37-84B cathedral (late version), shown on page 84.

81

Model 37-34

Frequency Coverage: 530-1720 kc
Power: 6-volt battery operated
Tubes Used: 5
Controls: 3; Tuning, Off-On, Volume
See Appendix II (10-5) for the tube diagram for Model 37-34.

Model 37-34 was offered in cabinets similar to Model 37-60. Please refer to Model 37-60 on page 82 for pictures of these.

Model 37-38

Frequency Coverage: Two bands.
    Band #1: 530-1720 kc
    Band #2: 2.3-7.4 mc
Power: Battery operated
Tubes Used: 6
Controls: 4; Tuning, Off-On, Band Switch, Volume
See Appendix II (10-6) for the tube diagram for Model 37-38.

Model 37-60

Frequency Coverage: Two bands.
    Band #1: 530-1720 kc
    Band #2: 2.3-7.4 mc
Power: AC; 115 volts, 50-60 cycles
Tubes Used: 5
Controls: 4; Tuning, Off-On/Tone, Band Switch, Volume
See Appendix II (10-7) for the tube diagram for Model 37-60.

Model 37-38F console.

Model 37-38J console.

Model 37-38B tombstone.

Model 37-60B cathedral.

Model 37-60B tombstone. *Photo by Ron Boucher.*

Model 37-61
Frequency Coverage: Two bands.
   Band #1: 530-1720 kc
   Band #2: 5.7-18.2 mc
Power: AC; 115 volts, 50-60 cycles
Tubes Used: 5
Controls: 4; Tuning, Off-On/Tone, Band Switch, Volume
See Appendix II (10-8) for the tube diagram for Model 37-61.

Model 37-61F console. *Photo by Edmund DeCann.*

Model 37-60F console.

Model 37-61B cathedral.

Model 37-62

Frequency Coverage: Two bands.
    Band #1: 530-1720 kc
    Band #2: 2.3-2.5 mc
Power: AC; 115 volts, 50-60 cycles
Tubes Used: 5
Controls: 3; Tuning, Off-On/Volume, Band Switch
See Appendix II (10-9) for the tube diagram for Model
37-62.

Model 37-62C was available in a table model cabinet.

Model 37-84

Frequency Coverage: 540-1700 kc
Power: AC; 115 volts, 50-60 cycles
Tubes Used: 4
Controls: 2; Tuning, Off-On/Volume
See Appendix II (10-10) for the tube diagram for Model
37-84.

Model 37-89

Frequency Coverage: Two bands.
    Band #1: 530-1650 kc
    Band #2: 1.5-3.7 mc
Power: AC; 115 volts, 50-60 cycles
Tubes Used: 6
Controls: 4; Tuning, Off-On/Tone, Band Switch, Volume
See Appendix II (10-11) for the tube diagram for Model
37-89.

Model 37-84B cathedral (early version). *Photo by Ron Boucher.*

Model 37-89B cathedral. *Photo by John Okolowicz.*

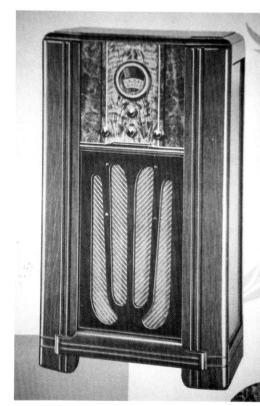

Model 37-84B cathedral (late version).

Model 37-89F console.

Model 37-93

Frequency Coverage: 530-1720 kc
Power: AC; 115 volts, 50-60 cycles
Tubes Used: 5
Controls: 2; Tuning, Off-On/Volume
See Appendix II (10-12) for the tube diagram for Model 37-93.

Model 37-116X De Luxe console.

Model 37-93B cathedral.

Models 37-600 & 37-602

Frequency Coverage: 530-1800 kc
Power:
    Model 37-600: AC; 115 volts, 50-60 cycles
    Model 37-602: 115 volts AC/DC
Tubes Used: see Appendix II
Controls: 2; Tuning, Off-On/Volume
See Appendix II (10-14) for the tube diagram for Model 37-600.

See Appendix II (10-15) for the tube diagram for Model 37-602.

Model 37-600C table model. *Photo by Doug Houston.*

Model 37-116

Frequency Coverage: Five bands.
Band #1: 530-1600 kc
Band #2: 1.58-4.75 mc
Band #3: 4.7-7.4 mc
Band #4: 7.35-11.6 mc
Band #5: 11.5-18.2 mc
Power: AC; 115 volts, 50-60 cycles
Tubes Used: 15
Controls: 6; Tuning, Off-On/Bass, Magnetic Tuning, Band Switch, Volume, Treble/Selectivity
Standard Model 37-116X uses a shadow meter. Model 37-116X De Luxe features Automatic Tuning. Both versions feature Magnetic Tuning (AFC), variable IF bandwidth, and Acoustic Clarifiers.
See Appendix II (10-13) for the tube diagram for Model 37-116X.

Model 37-602C table model.

Model 37-604

Frequency Coverage: Two bands.
   Band #1: 530-1750 kc
   Band #2: 6.0-18.0 mc
Power: 115 volts AC/DC
Tubes Used: 5
Controls: 3; Band Switch, Tuning, Off-On/Volume
See Appendix II (10-16) for the tube diagram for Model 37-604.

Model 37-610T table model. *Photo by Doug Houston.*

Model 37-604C table model. *Photo by Michael Prosise.*

Model 37-610

Frequency Coverage: Three bands.
   Band #1: 530-1720 kc
   Band #2: 2.3-7.4 mc
   Band #3: 7.35-22.0 mc
Power: AC; 115 volts, 50-60 cycles
Tubes Used: 5
Controls: 4; Tuning, Off-On/Tone, Band Switch, Volume

See Appendix II (10-17) for the tube diagram for Model 37-610.

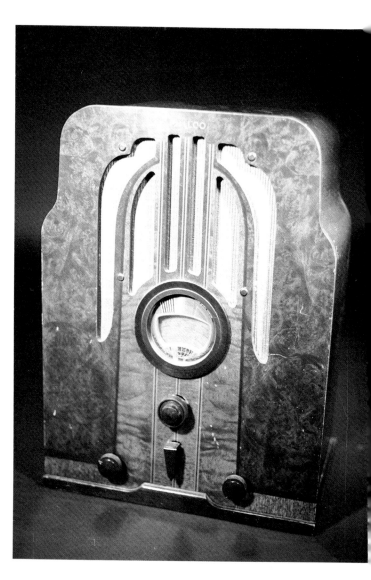

Model 37-610B tombstone. *Photo by Bob Schafbuch.*

Model 37-610J console.

Model 37-611

Frequency Coverage: Three bands.
    Band #1: 530-1720 kc
    Band #2: 2.3-7.4 mc
    Band #3: 7.35-22.0 mc
Power: 115 volts AC/DC
Tubes Used: 5
Controls: 4; Tuning, Off-On/Tone, Band Switch, Volume
See Appendix II (10-18) for the tube diagram for Model 37-611.

Models 37-611T table model and 37-611J console.

Models 37-611F console and 37-611B tombstone.

Model 37-620

Frequency Coverage: Three bands.
    Band #1: 530-1720 kc
    Band #2: 2.3-7.4 mc
    Band #3: 7.35-22.0 mc
Power: AC; 115 volts, 50-60 cycles
Tubes Used: 6
Controls: 4; Tuning; Off-On/Tone, Band Switch, Volume
See Appendix II (10-19) for the tube diagram for Model 37-620.

Model 37-620B tombstone. *Photo by Lewis Owens.*

87

Closed (left) and open (right) views of Model 37-620 Deluxe Radiobar. *Photos by Michael Prosise.*

Model 37-620T is in a cabinet similar to Model 38-620T table model, shown in Chapter Eleven on page 105.

Model 37-623

Frequency Coverage: Three bands.
    Band #1: 530-1720 kc
    Band #2: 2.3-7.4 mc
    Band #3: 7.35-22.0 mc
Power: Battery operated
Tubes Used: 6
Controls: 4; Tuning, Off-On/Tone, Band Switch, Volume
See Appendix II (10-20) for the tube diagram for Model 37-623.

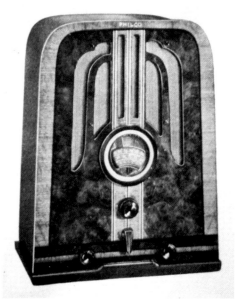

Model 37-620J console.

Model 37-623B tombstone.

Model 37-623J console.

Model 37-630X console. *Photo by Spencer Doggett.*

Model 37-624
Frequency Coverage: Three bands.
    Band #1: 530-1720 kc
    Band #2: 2.3-7.4 mc
    Band #3: 7.35-22.0 mc
Power: 6-volt battery operated
Tubes Used: 6
Controls: 4; Tuning, Off-On/Tone, Band Switch, Volume
See Appendix II (10-21) for the tube diagram for Model 37-624.

Model 37-624 was available in cabinets identical to Model 37-623, shown above.

Model 37-630
Frequency Coverage: Three bands.
    Band #1: 530-1720 kc
    Band #2: 2.3-7.4 mc
    Band #3: 7.35-22.0 mc
Power: AC; 115 volts, 50-60 cycles
Tubes Used: 6
Controls: 4; Tuning, Off-On/Tone, Band Switch, Volume
These models use shadow meters.
See Appendix II (10-22) for the tube diagram for Model 37-630.

Model 37-640
Frequency Coverage: Three bands.
    Band #1: 530-1720 kc
    Band #2: 2.3-7.4 mc
    Band #3: 7.35-22.0 mc
Power: AC; 115 volts, 50-60 cycles
Tubes Used: 7
Controls: 4; Tuning, Off-On/Tone, Band Switch, Volume
These models use shadow meters.
See Appendix II (10-23) for the tube diagram for Model 37-640.

Model 37-630T table model. *Photo by Jerry McKinney.*

Model 37-640B tombstone.

Model 37-640X console. *Photo by Doug Houston.*

Model 37-641
Frequency Coverage: Three bands.
  Band #1: 530-1720 kc
  Band #2: 2.3-7.4 mc
  Band #3: 7.35-22.0 mc
Power: 115 volts AC/DC
Tubes Used: 7
Controls: 4; Tuning, Off-On/Tone, Band Switch, Volume
These models use shadow meters.
See Appendix II (10-24) for the tube diagram for Model 37-641.

Model 37-641B tombstone.

Model 37-640MX console. *Photo by Doug Houston.*

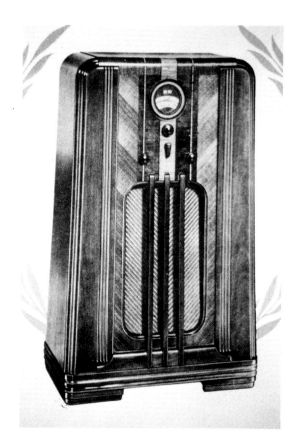

Model 37-641X console.

Model 37-641MX is in a cabinet identical to Model 37-640MX console, shown above.

Model 37-643

Frequency Coverage: Four bands.
    Band #1: 530-1600 kc
    Band #2: 1.58-4.8 mc
    Band #3: 4.7-11.6 mc
    Band #4: 11.5-18.2 mc
Power: Battery operated
Tubes Used: 7
Controls: 4; Tuning, Off-On/Tone, Band Switch, Volume
These models use shadow meters.
See Appendix II (10-25) for the tube diagram for Model 37-643.

Model 37-650

Frequency Coverage: Three bands.
    Band #1: 530-1720 kc
    Band #2: 5.7-11.6 mc
    Band #3: 11.5-18.2 mc
Power: AC; 115 volts, 50-60 cycles
Tubes Used: 8
Controls: 4; Tuning, Off-On/Tone, Band Switch, Volume
These models use shadow meters.
See Appendix II (10-26) for the tube diagram for Model 37-650.

Model 37-643B tombstone.

Model 37-650B tombstone.

Model 37-643X console.

Model 37-650X console. *Photo by Doug Houston.*

Model 37-660

Frequency Coverage: Four bands.
  Band #1: 530-1720 kc
  Band #2: 2.3-7.4 mc
  Band #3: 7.35-11.6 mc
  Band #4: 11.5-18.2 mc
Power: AC; 115 volts, 50-60 cycles
Tubes Used: 9
Controls: 4; Tuning, Off-On/Tone, Band Switch, Volume
These models use shadow meters.
 See Appendix II (10-27) for the tube diagram for Model 37-660.

Model 37-660B tombstone.

Model 37-665

Frequency Coverage: Four bands.
  Band #1: 530-1720 kc
  Band #2: 2.3-7.4 mc
  Band #3: 7.35-22.0 mc
  Band #4: 25.0-42.0 mc
Power: AC; 115 volts, 50-60 cycles
Tubes Used: 9
Controls: 4; Tuning, Off-On/Tone, Band Switch, Volume
These models use shadow meters.
 See Appendix II (10-28) for the tube diagram for Model 37-665.

Model 37-665B is in a cabinet identical to Model 37-660B tombstone, shown above.

Model 37-665X is in a cabinet identical to Model 37-660X console (in walnut only), shown at lower left.

Model 37-670B tombstone. *Photo by Wayne King.*

Model 37-670X console.

Model 37-670

Frequency Coverage: Five bands.
  Band #1: 530-1600 kc
  Band #2: 1.58-4.75 mc
  Band #3: 4.7-7.4 mc
  Band #4: 7.35-11.6 mc
  Band #5: 11.5-18.2 mc
Power: AC; 115 volts, 50-60 cycles
Tubes Used: 11
Controls: 4; Tuning, Off-On/Tone, Band Switch, Volume
These models use shadow meters.
 See Appendix II (10-29) for the tube diagram for Model 37-670.

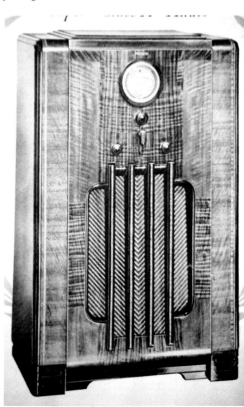

Model 37-660X console. This cabinet was available in walnut or mahogany.

Model 37-675

Frequency Coverage: Five bands.
  Band #1: 530-1600 kc
  Band #2: 1.58-4.75 mc
  Band #3: 4.7-7.4 mc
  Band #4: 7.35-11.6 mc
  Band #5: 11.5-18.2 mc
Power: AC; 115 volts, 50-60 cycles
Tubes Used: 12
Controls: 6; Tuning, Off-On/Bass, Magnetic Tuning, Band Switch,
Volume, Treble
Standard Model 37-675X uses a shadow meter. Model 37-675 De Luxe features Automatic Tuning. Both versions feature Magnetic Tuning (AFC).
See Appendix II (10-30) for the tube diagram for Model 37-675.

Model 37-675X De Luxe console.

Model 37-690

Frequency Coverage: Five bands.
  Band #1: 530-1600 kc
  Band #2: 1.58-4.75 mc
  Band #3: 4.7-7.4 mc
  Band #4: 7.35-11.6 mc
  Band #5: 11.5-18.2 mc
Power: AC; 115 volts, 50-60 cycles
Tubes Used: 20
Controls: 6; Tuning, Off-On/Bass, Magnetic Tuning, Band Switch, Volume, Treble/Selectivity
This model features Automatic Tuning, Magnetic Tuning (AFC), one woofer and two tweeters, Acoustic Clarifiers, and variable IF bandwidth
See Appendix II (10-31) for the tube diagram for the Model 37-690 main chassis.

See Appendix II (10-32) for the tube diagram for the Model 37-690 audio/power supply chassis.

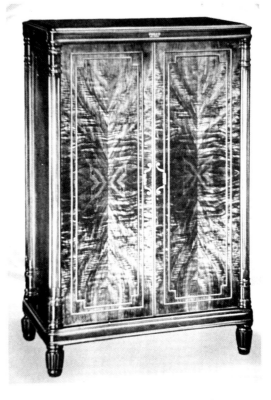

Closed (left) and open (right) views of Model 37-690X De Luxe console.

Model 37-2620

Frequency Coverage: Three bands.
  Band #1: 150-350 kc
  Band #2: 530-1720 kc
  Band #3: 5.7-18.0 mc
Power: AC; 115 volts, 50-60 cycles
Tubes Used: 6
Controls: 4; Tuning, Off-On/Tone, Band Switch, Volume
See Appendix II (10-33) for the tube diagram for Model 37-2620.

Model 37-2620B is in a cabinet identical to Model 37-620B tombstone, shown on page 87.

Model 37-2620J is in a cabinet identical to Model 37-620J console, shown on page 88.

Model 37-2650

Frequency Coverage: Four bands.
  Band #1: 150-350 kc
  Band #2: 530-1720 kc
  Band #3: 5.7-11.6 mc
  Band #4: 11.5-18.2 mc
Power: AC; 115 volts, 50-60 cycles
Tubes Used: 8
Controls: 4; Tuning, Off-On/Tone, Band Switch, Volume
These models use shadow meters.
See Appendix II (10-34) for the tube diagram for Model 37-2650.

Model 37-2650 was offered in the same cabinets as Model 37-650, shown on page 91.

Model 37-2670

Frequency Coverage: Five bands.
  Band #1: 150-350 kc
  Band #2: 530-1600 kc
  Band #3: 1.6-4.8 mc
  Band #4: 4.6-11.5 mc
  Band #5: 11.5-22.0 mc
Power: AC; 115 volts, 50-60 cycles
Tubes Used: 11
Controls: 4; Tuning, Off-On/Tone, Band Switch, Volume
These models use shadow meters.
See Appendix II (10-35) for the tube diagram for Model 37-2670.

Model 37-2670 was offered in the same cabinets as Model 37-670, shown on page 92.

"No squat, no stoop, no squint!" This was the big news from Philco this year. The slogan was used to describe their new line of console radios with inclined control panels that made the sets easier to operate.

Philco offered eight console models with the inclined control panel, along with more conventional consoles and tombstones. However, Philco was entering a new era, placing greater emphasis on compact table radios. Seventeen different models were offered in this style in 1938, while only twelve tombstone models were available.

Models 38-116 and 38-690, as well as the new models 38-1, 38-2 and 38-3, continued to feature the Automatic Tuning mechanism which Philco had introduced the previous season. In addition, the company now offered a new automatic tuning system which did not use Magnetic Tuning. It featured a large round dial with a tuning knob positioned on the edge of the dial. To tune in a station, the knob was moved around the circumference of the dial. Automatic tuning was accomplished by placing up to fifteen selector stops in the dial mechanism. In order to tune in a station automatically, the tuning knob was moved to the correct position on the dial, a cone clicked into place, and the station was then tuned in. Philco called it Cone-Centric Automatic Tuning, and it was available on models 38-4, 38-7 and 38-22.

Models 38-116 and 38-690 now offered improved magnetic tuning and push-pull beam power output, using two of the new 6L6G tubes.

Unit construction, with a subchassis mounted in the middle of the main chassis, was still used on many 1938 Philco models. However, several of the new models used more conventional chassis design. The models with unit construction were, and still are, difficult to service, especially the subchassis.

It was during the 1938 season that Philco made its ten millionth radio, which was a 38-116XX. Once this milestone had been reached, Philco offered a limited number of 38-116XX models with small brass plaques mounted on the cabinet under the band switch. The plaques read "Exact Replica of the Ten Millionth Philco" and also had the purchaser's name etched on it.

By now, Zenith was offering "Winchargers," small windmills which produced electricity and could be used to charge batteries, with their six-volt operated receivers. In 1938, Philco offered its own "Sky-Chargers," manufactured for Philco by the Parris-Dunn Corporation.

Also during the 1938 season, Philco re-entered the radio-phonograph business, after an absence of one year. New models included the 38-1-1PC, 38-3-3PC, 38-4-4PC, 38-5-9PS, 38-6-9PF and 38-1-116PC. Each model used automatic record changers made by Capehart, which at the time was one of the best known manufacturers of automatic phonographs.

Total Philco radio sales in 1938 dropped to one million units. On the first of May, as Philco was preparing to introduce its 1939 line of radio receivers, the company's employees went on strike. The strike had a lot to do with the drop in Philco's sales for the calendar year 1938. Meanwhile, the company was preparing for a change in direction.

No Squat, No Stoop, No Squint

What new idea could a radio manufacturer offer the public, after they had introduced (or adopted) innovations such as four point tone control, cathedral radios, chairside radios, the Inclined Sounding Board, 6.3 volt tubes, Super Class A audio, high fidelity, Acoustic Clarifiers, Automatic Tuning and Magnetic Tuning?

Philco's answer for 1938 was the inclined control panel. Now it was no longer necessary to bend over or squat down to see the dial while operating the set. Because of this, Philco coined the phrase "No Squat, No Stoop, No Squint" to describe the new consoles with inclined control panels.

Philco's 1938 catalogs and brochures included several drawings of people in squatting positions or bent over, tuning a conventional console radio, in order to illustrate the advantages of the inclined control panel. Most of the drawings were of women. This was no accident, as Philco's advertising campaign promoting the new design was deliberately targeted toward the ladies. For example, the 1938 dealer catalog mentions, "Here's a radio that a woman can tune with *ease* and with *grace*..."

Philco's new slogan was parodied in two 1938 Warner Bros./ Vitaphone cartoons. In *Cinderella Meets Fella,* the Cinderella character runs over to her radio (which is a rendition of a 38-116XX) and, while giggling, points to it and says, "No squat, no squint, no stoop!" And in *Johnny Smith and Poker-Huntas,* the Indian chief's name is Chief No Squat No Stoop No Squint.

Although Philco's "No Squat, No Stoop, No Squint" slogan may have been joked about, the inclined control panel would eventually be copied by several manufacturers. So perhaps Philco had the last laugh after all.

Model 38-1
Frequency Coverage: Three bands.
    Band #1: 530-1720 kc
    Band #2: 2.3-7.4 mc
    Band #3: 7.35-22.0 mc
Power: AC; 115 volts, 50-60 cycles
Tubes Used: 12
Controls: 5; Tuning, Off-On/Tone, Band Switch, Magnetic Tuning,
Volume
This model features Automatic Tuning and Magnetic Tuning (AFC)
See Appendix II (11-1) for the tube diagram for Model 38-1.

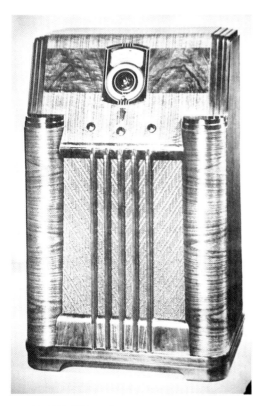

Model 38-1XX console.

Model 38-1-1PC is a radio-phonograph console using the 38-1 chassis.

Model 38-2
Frequency Coverage: Three bands.
    Band #1: 530-1720 kc
    Band #2: 2.3-7.4 mc
    Band #3: 7.35-22.0 mc
Power: AC; 115 volts, 50-60 cycles
Tubes Used: 11
Controls: 5; Tuning, Off-On/Tone, Band Switch, Magnetic Tuning, Volume
This model features Automatic Tuning and Magnetic Tuning (AFC)
See Appendix II (11-2) for the tube diagram for Model 38-2.

Model 38-2XX console.

Model 38-3
Frequency Coverage: Three bands.
    Band #1: 530-1720 kc
    Band #2: 2.3-7.4 mc
    Band #3: 7.35-22.0 mc
Power: AC; 115 volts, 50-60 cycles
Tubes Used: 9
Controls: 5; Tuning, Off-On/Tone, Band Switch, Magnetic Tuning, Volume
This model features Automatic Tuning and Magnetic Tuning (AFC)
See Appendix II (11-3) for the tube diagram for Model 38-3.

Model 38-3-3PC Radio-Phonograph. *Photo by Doug Houston.*

Model 38-3XX console.

Model 38-4
Frequency Coverage: Two bands.
Band #1: 540-1720 kc
Band #2: 5.7-18.2 mc
Power: AC; 115 volts, 50-60 cycles
Tubes Used: 8
Controls: 4; Tuning, Off-On/Tone, Band Switch, Volume
This model features Automatic Tuning
See Appendix II (11-4) for the tube diagram for Model 38-4.

Model 38-4-4PC is a radio-phonograph console using the 38-4 chassis.

Model 38-5
Frequency Coverage: Two bands.
Band #1: 540-1720 kc
Band #2: 5.7-18.2 mc

Power: AC; 115 volts, 50-60 cycles
Tubes Used: 8
Controls: 4; Tuning, Off-On/Tone, Band Switch, Volume
See Appendix II (11-5) for the tube diagram for Model 38-5.

Model 38-5X console.

Model 38-5B is in a cabinet identical to Model 38-665B tombstone, shown on page 107.

Model 38-7
Frequency Coverage: Two bands.
Band #1: 530-1720 kc
Band #2: 5.7-18.0 mc
Power: AC; 115 volts, 50-60 cycles
Tubes Used: 6

Controls: 4; Tuning, Off-On/Tone, Band Switch, Volume
These models feature Automatic Tuning
See Appendix II (11-6) for the tube diagram for Model 38-7.

Model 38-7T table model. *Photo by Doug Houston.*

Model 38-4XX console.

Model 38-7XX console.

Model 38-7CS chairside set. *Photo by Doug Houston.*

Models 38-8 & 38-9
(both used the same chassis)
Frequency Coverage: Two bands.
   Band #1: 530-1720 kc
   Band #2: 5.7-18.0 mc
Power: AC; 115 volts, 50-60 cycles
Tubes Used: 6
Controls: 4; Tuning, Off-On/Tone, Band Switch, Volume
Model 38-8X uses a shadow meter.
See Appendix II (11-7) for the tube diagram for Models 38-8 & 38-9.

Model 38-8X console. *Photo by Edmund DeCann.*

Model 38-9T table model. *Photo by Spencer Doggett.*

Model 38-9 Radio-Phonograph. *Photo by Doug Houston.*

Model 38-10
Frequency Coverage: Two bands.
    Band #1: 540-1720 kc
    Band #2: 5.7-18.0 mc
Power: AC; 115 volts, 50-60 cycles
Tubes Used: 5
Controls: 4; Tuning, Off-On/Tone, Band Switch, Volume
See Appendix II (11-8) for the tube diagram for Model 38-10.

Model 38-9K console.

Model 38-10T table model.

Model 38-10F console.

Closed (top) and open (bottom) views of Model 38-10 chairside Radiobar. *Photos by Michael Prosise.*

Model 38-12
Frequency Coverage: 540-1720 kc
Power: AC; 115 volts, 50-60 cycles
Tubes Used: 5
Controls: 2; Off-On/Volume, Tuning
See Appendix II (11-9) for the tube diagram for Model 38-12.

Model 38-12C table model. *Photo by Doug Houston.*

Model 38-12CI is similar to Model 38-12C, but with an ivory finish.

Model 38-12T is in a cabinet similar to Model 38-15T table model, shown on page 100.

Model 38-12 was also offered in brown and ivory bakelite table model cabinets.

Model 38-14

Frequency Coverage: Two bands.
  Band #1: 530-1720 kc
  Band #2: 2.3-7.4 mc
Power: 115 volts AC/DC
Tubes Used: 5
Controls: 3; Off-On/Volume, Band Switch, Tuning
See Appendix II (11-10) for the tube diagram for Model 38-14.

Model 38-14T is in a cabinet identical to Model 38-15T table model, shown below.

Model 38-14CS is a chairside set using the 38-14 chassis.

Model 38-14 was also offered in brown and ivory bakelite table model cabinets.

Model 38-15

Frequency Coverage: Two bands.
  Band #1: 540-1720 kc
  Band #2: 5.7-18.0 mc
Power: AC; 115 volts, 50-60 cycles
Tubes Used: 5
Controls: 3; Off-On/Volume, Band Switch, Tuning
See Appendix II (11-11) for the tube diagram for Model 38-15.

Model 38-15T table model. *Photo by Michael Prosise.*

Unusual model 38-15 table model, which despite the two speaker openings, uses only one speaker. *Photo by Michael Prosise.*

Model 38-15CS is a chairside set using the 38-15 chassis.

Model 38-15 was also offered in brown and ivory bakelite table model cabinets.

Model 38-22

Frequency Coverage: Two bands.
  Band #1: 530-1720 kc
  Band #2: 5.7-18.0 mc
Power: 115 volts AC/DC
Tubes Used: 6
Controls: 4; Tuning, Off-On/Tone, Band Switch, Volume
These models feature Automatic Tuning.
See Appendix II (11-12) for the tube diagram for Model 38-22.

Model 38-22XX console.

Model 38-22T is in a cabinet identical to Model 38-7T table model, shown on page 96.

Model 38-22CS is in a cabinet identical to Model 38-7CS chairside set, shown on page 97.

Model 38-23

Frequency Coverage: Two bands.
  Band #1: 530-1720 kc
  Band #2: 5.7-18.0 mc
Power: 115 volts AC/DC
Tubes Used: 6
Controls: 4; Tuning, Off-On/Tone, Band Switch, Volume
See Appendix II (11-13) for the tube diagram for Model 38-23.

Model 38-23X console.

Model 38-23T is in a cabinet identical to Model 38-10T table model, shown on page 98.

Model 38-23K is in a cabinet identical to Model 38-9K console, shown on page 98.

Model 38-33

Frequency Coverage: 530-1720 kc
Power: Battery operated
Tubes Used: 5
Controls: 2; Tuning, Off-On/Volume
See Appendix II (11-14) for the tube diagram for Model 38-33.

Model 38-33F console.

Model 38-34

Frequency Coverage: 530-1720 kc
Power: 6-volt battery operated
Tubes Used: 5
Controls: 2; Tuning, Off-On/Volume
See Appendix II (11-15) for the tube diagram for Model 38-34.

Model 38-34 was available in tombstone and console cabinets.

Model 38-35

Frequency Coverage: 530-1720 kc
Power: 115 volt, 60 cycle AC or 6-volt battery
Tubes Used: 5
Controls: 2; Tuning, Off-On/Volume
See Appendix II (11-16) for the tube diagram for Model 38-35.

Model 38-35 was available in tombstone and console cabinets.

Model 38-38

Frequency Coverage: Two bands.
    Band #1: 530-1720 kc
    Band #2: 5.7-18.0 mc
Power: Battery operated
Tubes Used: 6
Controls: 4; Tuning, Off-On/Tone, Band Switch, Volume
See Appendix II (11-17) for the tube diagram for Model 38-38.

Model 38-38T table model.

Model 38-33B tombstone.

101

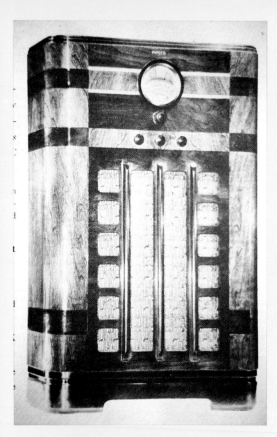

Model 38-38K console.

Models 38-39 & 38-40

Freqeuncy Coverage: Two bands.
    Band #1: 530-1720 kc
    Band #2: 5.7-18.0 mc
Power:
    Model 38-39: 6-volt battery operated
    Model 38-40: 115 volt, 60 cycle AC or 6-volt battery
Tubes Used: 6
Controls: 4; Tuning, Off-On/Tone, Band Switch, Volume
See Appendix II (11-18) for the tube diagram for Model 38-39
See Appendix II (11-19) for the tube diagram for Model 38-40.

Models 38-39 and 38-40 were both offered in cabinets identical to Model 38-38, shown at left and on page 101.

Model 38-60

Frequency Coverage: Two bands.
    Band #1: 530-1720 kc
    Band #2: 2.3-7.4 mc
Power: AC; 115 volts, 50-60 cycles
Tubes Used: 5
Controls: 4; Tuning, Off-On/Tone, Band Switch, Volume

See Appendix II (11-20) for the tube diagram for Model 38-60

Model 38-60B tombstone.

Model 38-38X console.

Model 38-60F console.

Model 38-62
Frequency Coverage: Two bands.
    Band #1: 530-1720 kc
    Band #2: 2.3-2.5 mc
Power: AC; 115 volts, 50-60 cycles
Tubes Used: 5
Controls: 3; Tuning, Off-On/Volume, Band Switch
See Appendix II (11-21) for the tube diagram for Model
38-62

Model 38-62C table model. *Photo by Doug Houston.*

Model 38-62F console.

Model 38-89
Frequency Coverage: Two bands.
    Band #1: 530-1650 kc
    Band #2: 1.5-3.7 mc
Power: AC; 115 volts, 50-60 cycles
Tubes Used: 6
Controls: 4; Tuning, Off-On/Tone, Band Switch, Volume
See Appendix II (11-22) for the tube diagram for Model
38-89

Model 38-89B tombstone.

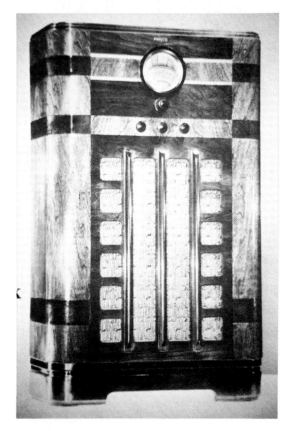

Model 38-89K console.

Model 38-93

Frequency Coverage: 530-1720 kc
Power: AC; 115 volts, 50-60 cycles
Tubes Used: 5
Controls: 2; Tuning, Off-On/Volume
See Appendix II (11-23) for the tube diagram for Model 38-93

Model 38-93B tombstone (early version).

Model 38-93B tombstone (late version). *Photo by Ron Boucher.*

Model 38-116

Frequency Coverage: Five bands.
  Band #1: 530-1600 kc
  Band #2: 1.58-4.75 mc
  Band #3: 4.7-7.4 mc
  Band #4: 7.35-11.6 mc

Band #5: 11.5-18.2 mc
Power: AC; 115 volts, 50-60 cycles
Tubes Used: 15
Controls: 6; Tuning, Off-On/Bass, Magnetic Tuning, Band Switch, Volume, Treble/Selectivity
This model features Automatic Tuning, Magnetic Tuning (AFC), variable IF bandwidth and Acoustic Clarifiers
See Appendix II (11-24) for the tube diagram for Model 38-116

Model 38-116XX console. *Photo by John Okolowicz.*

Model 38-1-116PC is a radio-phonograph console using the 38-116 chassis.

An "Exact Replica of the Ten Millionth Philco" model 38-116XX.

Close-up of the brass plaque placed on special "Ten Millionth Philco" receivers.

Model 38-610

Frequency Coverage: Three bands.
   Band #1: 530-1720 kc
   Band #2: 2.3-7.4 mc
   Band #3: 7.35-22.0 mc
Power: AC; 115 volts, 50-60 cycles
Tubes Used: 5
Controls: 4; Tuning, Off-On/Tone, Band Switch, Volume

Model 38-610B tombstone.

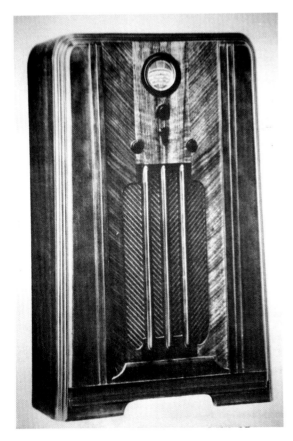

Model 38-610J console.

Model 38-620

Frequency Coverage: Three bands.
   Band #1: 530-1720 kc
   Band #2: 2.3-7.4 mc
   Band #3: 7.35-22.0 mc
Power: AC; 115 volts, 50-60 cycles
Tubes Used: 6
Controls: 4; Tuning, Off-On/Tone, Band Switch, Volume
See Appendix II (11-26) for the tube diagram for Model 38-620

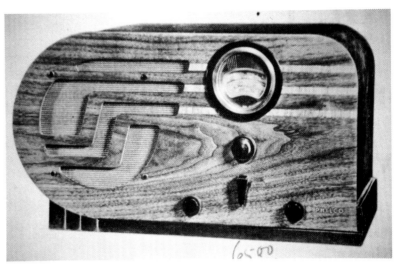

Model 38-620T table model.

Model 38-623

Frequency Coverage: Three bands.
   Band #1: 530-1720 kc
   Band #2: 2.3-7.4 mc
   Band #3: 7.35-22.0 mc
Power: Battery operated
Tubes Used: 6
Controls: 4; Tuning, Off-On/Tone, Band Switch, Volume
See Appendix II (11-27) for the tube diagram for Model 38-623

Model 38-623T table model.

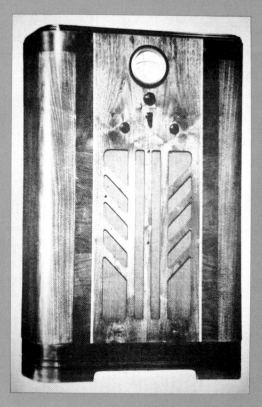

Model 38-623K console.

Model 38-624
Frequency Coverage: Three bands.
    Band #1: 530-1720 kc
    Band #2: 2.3-7.4 mc
    Band #3: 7.35-22.0 mc
Power: 6-volt battery operated
Tubes Used: 6
Controls: 4; Tuning, Off-On/Tone, Band Switch, Volume
See Appendix II (11-28) for the tube diagram for Model 38-624.

Model 38-624 was offered in cabinets identical to Model 38-623, shown above and on page 105.

Model 38-630
Frequency Coverage: Three bands.
    Band #1: 530-1720 kc
    Band #2: 2.3-7.4 mc
    Band #3: 7.35-22.0 mc
Power: AC; 115 volts, 50-60 cycles
Tubes Used: 6
Controls: 4; Tuning, Off-On/Tone, Band Switch, Volume
This model uses a shadow meter.
See Appendix II (11-29) for the tube diagram for Model 38-630.

Model 38-630K console.

Model 38-643
Frequency Coverage: Four bands.
    Band #1: 530-1600 kc
    Band #2: 1.58-4.8 mc
    Band #3: 4.7-11.6 mc
    Band #4: 11.5-18.2 mc
Power: Battery operated
Tubes Used: 7
Controls: 4; Tuning, Off-On/Tone, Band Switch, Volume
These models use shadow meters.
See Appendix II (11-30) for the tube diagram for Model 38-643.

Model 38-643B tombstone.

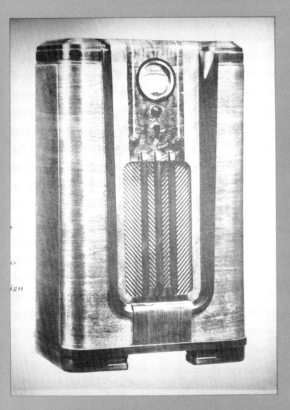

Model 38-643X console.

Model 38-665

Frequency Coverage: Four bands.
  Band #1: 530-1720 kc
  Band #2: 2.3-7.4 mc
  Band #3: 7.35-22.0 mc
  Band #4: 25.0-42.0 mc
Power: AC; 115 volts, 50-60 cycles
Tubes Used: 9
Controls: 4; Tuning, Off-On/Tone, Band Switch, Volume
These models use shadow meters.
 See Appendix II (11-31) for the tube diagram for Model 38-665.

Model 38-665B tombstone.

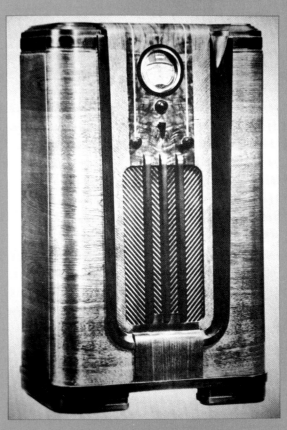

Model 38-665X console.

Model 38-690

Frequency Coverage: Five bands.
  Band #1: 530-1600 kc
  Band #2: 1.58-4.75 mc
  Band #3: 4.7-7.4 mc
  Band #4: 7.35-11.6 mc
  Band #5: 11.5-18.2 mc
Power: AC; 115 volts, 50-60 cycles
Tubes Used: 20
Controls: 6; Tuning, Off-On/Bass, Magnetic Tuning, Band Switch,
Volume, Treble/Selectivity
This model features Automatic Tuning, Magnetic Tuning (AFC), one woofer and two tweeters, Acoustic Clarifiers, and variable IF bandwidth

See Appendix II (11-32) for the tube diagram for Model 38-690, the main chassis.

See Appendix II (11-33) for the tube diagram for Model 38-690, the audio/power supply chassis.

Closed (left) and open (right) views of Model 38-690XX console. *Photos by Michael Prosise.*

Models 38-2620 & 38-2630

(both used very similar chasses)
Frequency Coverage: Three bands.
    Band #1: 150-395 kc
    Band #2: 530-1720 kc
    Band #3: 5.7-18.0 mc
Power: AC; 115 volts, 50-60 cycles
Tubes Used: 6
Controls: 4; Tuning, Off-On/Tone, Band Switch, Volume
Model 38-2630K uses a shadow meter.
See Appendix II (11-34) for the tube diagram for Models 38-2620 & 38-2630.

Model 38-2620T is in a cabinet identical to Model 38-620T table model, shown on page 105.
Model 38-2630K is in a cabinet identical to Model 38-630K console, shown on page 106.

Model 38-2650

Frequency Coverage: Four bands.
    Band #1: 150-395 kc
    Band #2: 530-1720 kc
    Band #3: 5.7-11.6 mc
    Band #4: 11.5-18.2 mc
Power: AC; 115 volts, 50-60 cycles
Tubes Used: 8
Controls: 4; Tuning, Off-On/Tone, Band Switch, Volume
These models use shadow meters.
See Appendix II (11-35) for the tube diagram for Model 38-2650.

Model 38-2650 was offered in the same cabinets as Model 38-665, shown on page 107.

Model 38-2670

Frequency Coverage: Five bands.
    Band #1: 150-395 kc
    Band #2: 530-1600 kc
    Band #3: 1.6-4.8 mc
    Band #4: 4.6-11.5 mc
    Band #5: 11.5-22.0 mc
Power: AC; 115 volts, 50-60 cycles
Tubes Used: 11
Controls: 4; Tuning, Off-On/Tone, Band Switch, Volume
These models use shadow meters.
See Appendix II (11-36) for the tube diagram for Model 38-2670.

Model 38-2670X console.

Model 38-2670B tombstone. *Photo by Doug Houston.*

Philco workers had been on strike for one month in June 1938 when the company went ahead and introduced its new 1939 line. The company began to change its direction by offering not only radios and radio-phonographs, but also an air conditioner. The Cool-Wave air conditioners were made by York for Philco. Over the next several years, Philco would continue to diversify into non-electronic items.

Meanwhile, the new Philco radio line was not what it used to be, as most new models were completely new and different from previous designs. Tombstone models had all but disappeared (only three tombstones were offered, all of which were farm sets). Many new models used slide rule dials. Philco's previous automatic tuning systems were discarded in favor of pushbuttons for favorite standard broadcast stations.

Philco initially offered fewer models for 1939 than the previous season, as sixty radios, phonographs, and radio/phonographs were available. The company also offered more table models (twenty-two) than were available in 1938.

Philco had been a pioneer in high fidelity reproduction since 1932. But now, in the new 1939 models, Philco's high fidelity advances were tossed aside. The reason for this is not really known. Was it because Edwin H. Armstrong's latest invention, frequency modulation or FM - which would offer far greater fidelity with an extra added bonus of reducing static to a minimum - was on the horizon?

The trend toward lower quality sets was embodied in the new flagship model, the 39-116RX. It was housed in a large, impressive cabinet, and still covered 540 kc to 18 mc, but now in three bands instead of five, as last season's 38-690 and 38-116 had done. But the new model was no 38-690, or even a 38-116, as it offered no fidelity-selectivity control, Acoustic Clarifiers, or even separate bass and treble controls (only a simple tone control was used in the 39-116RX).

The set needed something new to make it desirable to persons who were willing to purchase a high priced radio, and Philco provided a new innovation for it, as well as the smaller 39-55RX: Mystery Control. This new feature consisted of a wireless remote control box with a type 30 tube and necessary batteries inside. Outside, it had a dial somewhat like a telephone. With the control box, a set owner could select one of eight preset standard broadcast stations, turn the volume up or down, and turn the radio off.

Philco had sold automobile radios under the Transitone name since 1930. For the 1939 season, Philco introduced two new Transitone sets for the home. These were small table models, which received the standard broadcast band only. The cheapest model, the TH-l, was a TRF receiver which used four tubes. The TH-3 was a five tube superheterodyne. The new Transitone line was intended to get customers to purchase second or even third radios for their homes which already had one or two sets.

An interesting model is the 39-50RX. This single-band console was equipped with a wired remote control, and had no dial or controls on the console cabinet. All controls (off-on/volume and eight pushbuttons for selecting favorite standard broadcast stations) were on the remote control unit. This set is not documented in *Rider's Perpetual Troubleshooter's Manual,* the main source of service information for radio receivers of the period (other than factory service data). Philco's 1946 parts catalog, which lists all Philco receivers produced through 1942, does not mention it either. By October 1939, the set was no longer mentioned in Philco's price lists. Whether or not any were actually made is not known.

In September, the strike at Philco ended, as a settlement was finally reached between the company and the union representing Philco's employees. In November, Fairbanks-Morse & Company sold its refrigerator division to Philco. The company was continuing to diversify as it would soon be in the refrigerator business.

Philco's Transitone line was expanded in January 1939 with the addition of five new models. Meanwhile, Philco's regular line added thirteen new models while discontinuing nine others. In addition, the company was now offering refrigerators as well as air conditioners.

In May, James M. Skinner was named Philco's new Chairman of the Board, succeeding Edward Davis. Skinner had been with the company since 1911, and was largely responsible for Philco's success. However, before the year was over, Skinner would leave the company which he had guided to the top of the radio market.

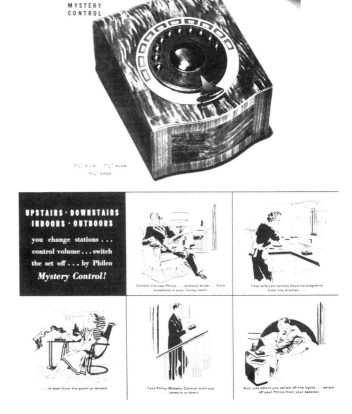

39-116RX Mystery Control, from the 1939 Philco dealer catalog. A similar remote control unit was used with all Philco sets with wireless remote control.

The 39-116 and Mystery Control

The 1939 Philco radio line was headed by a longer, lower, more streamlined model 116. Some variation of a 16 or 116 had come to symbolize Philco since the 1933-34 season, despite the fact that they had made better models such as the 200X, 201X, 680X, 37-690X and 38-690XX.

Since the 1937 season, the 116 had embodied most of the best features that Philco had to offer, such as variable IF bandwidth, separate bass and treble controls, automatic tuning, magnetic tuning, large speakers, Acoustic Clarifiers, and frequency coverage which included standard broadcast and shortwave frequencies up to 18 megacycles. Two stages of IF amplification had been part of 116 and 16 models since the 1933-34 season.

However, the 1939 model 116 was vastly different. When compared to previous top end Philco models, the sales pitch for model 39-116RX in their 1939 radio brochure sounds like a joke: "Philco 116RX brings you every new improvement in tone, performance, and beauty." The new model still used the Inclined Sounding Board, and covered the same frequencies as the 38-116, but only used three bands instead of five so the shortwave frequencies were crowded closer together. The set had only one stage of IF amplification and a single tone control. Philco's high fidelity innovations such as variable IF bandwidth and Acoustic Clarifiers were now gone.

For the new 116 and the smaller 39-55RX, a new novelty was devised by Philco in hopes of catching the buying public's fancy - Mystery Control. The separate wireless remote allowed a Mystery Control set owner to control some functions of the set without getting up from the easy chair. It is worthy of note that Philco claimed in its 1939 dealer catalog that Mystery Control was "not a radio beam," when in fact it was. When the remote unit was operated, it sent a series of pulses, via radio waves operating between 355 and 395 kc (depending on the individual receiver) to the radio. A special antenna in the radio picked up the pulses and sent them into a portion of the set which amplified and processed the pulses. It was actually a radio within a radio. The number of pulses told the radio whether to select a certain preset standard broadcast station, turn the volume up or down, or turn the set off.

Mystery Control models were designed so that the remote control could be used across the room or even upstairs. A sensitivity control was provided at the back of the radio to control how far away the remote could be used. Each Mystery Control radio and its matching remote were set to one of five control frequencies, in an effort to avoid the possibility of one set owner's remote controlling his or her neighbor's Mystery Control radio. However, as the sets became popular, the inevitable happened; quite frequently, one person's Mystery Control remote would, indeed, control the neighbor's set.

The Mystery Control remote unit used one tube, and five more tubes in the set were used for the wireless remote control. When the tubes used for Mystery Control are not counted, the 39-116 is basically an eight tube receiver. In comparison, the 38-116 uses fifteen tubes. When the tubes used for magnetic tuning are not counted, the 38-116 is still a twelve tube receiver. However, the 1939 model sold for $198, only slightly less than the 38-116's price of $200.

While the 39-116 cannot be compared to the 38-116, Mystery Control did prove to be popular, as it provided a new and novel means of operating a radio by remote control.

Models 39-6 & 39-7
(both used very similar chasses)
Frequency Coverage: 530-1720 kc
Power: AC; 115 volts, 50-60 cycles
Tubes Used: 5
Controls: 2; Off-On/Volume, Tuning
Model 39-7 features six pushbuttons for automatic tuning
See Appendix II (12-1) for the tube diagram for Models 39-6 & 39-7.

Model 39-6C table model. *Photo by Michael Prosise.*

Philco's first air conditioner, the "Cool-Wave," manufactured by York.

Model 39-6CI table model in ivory finish. *Photo by Gary Schieffer.*

Model 39-7C table model. *Photo by Edmund DeCann.*

Model 39-8T table model. *Photo by Doug Houston.*

Model 39-10
This table model semi-automatic record player is designed to broadcast to a nearby radio, whereby records could be played through the radio.

Model 39-7T table model.

Model 39-8
Frequency Coverage: 530-1720 kc
Power: 115 volts AC/DC
Tubes Used: 5
Controls: 2; Off-On/Volume, Tuning
See Appendix II (12-2) for the tube diagram for Model 39-8.

Model 39-10RP record player.

Model 39-12
Frequency Coverage: 540-1720 kc
Power: AC; 115 volts, 50-60 cycles
Tubes Used: 5
Controls: 2; Off-On/Volume, Tuning
See Appendix II (12-3) for the tube diagram for Model 39-12.

Model 39-12CB bakelite table model.

Model 39-12T table model.

Model 39-12CBI bakelite table model with ivory finish.

Model 39-12TP table model Radio-Phonograph.

Model 39-14

Frequency Coverage: Two bands.
    Band #1: 530-1720 kc
    Band #2: 2.3-7.4 mc
Power: 115 volts AC/DC
Tubes Used: 5
Controls: 3; Off-On/Volume, Band Switch, Tuning
See Appendix II (12-4) for the tube diagram for Model 39-14.

Model 39-14CB is in a cabinet similar to Model 39-12CB bakelite table model, shown above left.

Model 39-14CBI is in a cabinet similar to Model 39-12CBI bakelite table model with ivory finish, shown at left.

Model 39-14T is in a cabinet similar to Model 39-12T table model, shown above.

Model 39-15

Frequency Coverage: Two bands.
    Band #1: 540-1720 kc
    Band #2: 5.7-18.0 mc
Power: AC; 115 volts, 50-60 cycles
Tubes Used: 5
Controls: 3; Off-On/Volume, Band Switch, Tuning
See Appendix II (12-5) for the tube diagram for Model 39-15.

Model 39-15CB is in a cabinet similar to Model 39-12CB bakelite table model, shown at upper left.

Model 39-15CBI is in a cabinet similar to Model 39-12CBI bakelite table model with ivory finish, shown at left.

Model 39-15T is in a cabinet similar to Model 39-12T table model, shown above.

Models 39-17 & 39-117

(both used very similar chasses)
Frequency Coverage: 540-1720 kc
Power: AC; 115 volts, 50-60 cycles
Tubes Used: 5
Controls: 2; Off-On/Volume, Tuning
These models feature six pushbuttons for automatic tuning
See Appendix II (12-6) for the tube diagram for Models 39-17 & 39-117.

Model 39-17T table model.

Model 39-19PA Radio-Phonograph.

Model 39-17F console.

Model 39-117 was offered in the same cabinets as Model 39-17.

Models 39-18 & 39-118

(both used very similar chasses)
Frequency Coverage: 540-1720 kc
Power: 115 volts AC/DC
Tubes Used: 5, with one ballast
Controls: 2; Off-On/Volume, Tuning
These models feature six pushbuttons for automatic tuning
See Appendix II (12-7) for the tube diagram for Models 39-18 & 39-118.

Models 39-18 & 39-118 were offered in the same cabinets as Model 39-17, shown above.

Models 39-19 & 39-119

(both used very similar chasses)
Frequency Coverage: Two bands.
   Band #1: 540-1720 kc
   Band #2: 5.5-19.0 mc
Power: AC; 115 volts, 50-60 cycles
Tubes Used: 5
Controls: 3; Off-On/Volume, Tuning, Band Switch
These models feature six pushbuttons for automatic tuning
See Appendix II (12-8) for the tube diagram for Models 39-19 & 39-119.

Model 39-19PF Radio-Phonograph.

Model 39-19PCS chairside Radio-Phonograph.

Model 39-25T table model. *Photo by John Miller.*

Model 39-19TP table model Radio-Phonograph.

Model 39-19 was also available in cabinets identical to Model 39-17, shown on page 113.

Model 39-119 was offered in cabinets identical to Model 39-17, shown on page 113.

Model 39-25

Frequency Coverage: Two bands.
    Band #1: 540-1720 kc
    Band #2: 4.9-18.0 mc
Power: AC; 115 volts, 50-60 cycles
Tubes Used: 5
Controls: 4; Off-On/Tone, Volume, Band Switch, Tuning
These models feature eight pushbuttons for automatic tuning
See Appendix II (12-9) for the tube diagram for Model 39-25.

Model 39-25XF console.

Models 39-30, 39-31 & 39-35

(all used the same chassis)
Frequency Coverage: Two bands.
    Band #1: 540-1720 kc
    Band #2: 4.9-18.0 mc
Power: AC; 115 volts, 50-60 cycles
Tubes Used: 6
Controls: 4; Off-On/Tone, Volume, Band Switch, Tuning
These models feature eight pushbuttons for automatic tuning
See Appendix II (12-10) for the tube diagram for Models 39-30, 39-31 & 39-35.

Model 39-30T table model.

Model 39-30PCX Radio-Phonograph.

Model 39-31XF console.

Model 39-35XX console.

Model 39-3-31PA is a radio-phonograph console using the 39-31 chassis.

Model 39-31 was also available in an "XK" console cabinet.

Model 39-3-35PC is a radio-phonograph console using the 39-35 chassis.

Model 39-36
Frequency Coverage: Two bands.
    Band #1: 540-1720 kc
    Band #2: 5.0-18.0 mc
Power: AC; 115 volts, 50-60 cycles
Tubes Used: 6
Controls: 4; Tone, Volume, Tuning, Band Switch
This model features eight pushbuttons for automatic tuning
 See Appendix II (12-11) for the tube diagram for Model 39-36.

Model 39-36XX console.

Model 39-40

Frequency Coverage: Two bands.
    Band #1: 540-1720 kc
    Band #2: 5.8-18.0 mc
Power: AC; 115 volts, 50-60 cycles
Tubes Used: 8
Controls: 4; Tone, Volume, Tuning, Band Switch
These models feature eight pushbuttons for automatic tuning
See Appendix II (12-12) for the tube diagram for Model 39-40.

Model 39-40PCX Radio-Phonograph.

Model 39-2-40PC is a radio-phonograph console using the 39-40 chassis.

Model 39-45

Frequency Coverage: Three bands.
    Band #1: 540-1720 kc
    Band #2: 1.7-5.9 mc
    Band #3: 5.8-18.0 mc
Power: AC; 115 volts, 50-60 cycles
Tubes Used: 9
Controls: 4; Tone, Volume, Tuning, Band Switch
This model features eight pushbuttons for automatic tuning
See Appendix II (12-13) for the tube diagram for Model 39-45.

Model 39-45XX console.

Model 39-40XX console.

Model 39-50

Frequency Coverage: Eight pushbuttons which can be set to eight different frequencies in the standard broadcast (540-1600 kc) band
Power: AC; 115 volts, 50-60 cycles
Tubes Used: 7
Controls: 1; Off-On/Volume (on remote control)
This model features eight pushbuttons (on remote control) for automatic tuning; there are no controls on the console cabinet

Model 39-50RX console.

Remote control unit for Model 39-50RX.

Model 39-55

Frequency Coverage: 540-1720 kc
Power: AC; 115 volts, 50-60 cycles
Tubes Used: 11
Controls: 4; Tone, Off-On/Volume, Tuning, Standard/Remote Control
This model features Mystery (wireless remote) Control

See Appendix II (12-14) for the tube diagram for Model 39-55.

Model 39-55RX console. This uses a remote control box identical to that used with Models 39-116RX and 39-116PCX.

Model 39-70

Frequency Coverage: 530-1720 kc
Power: Battery operated
Tubes Used: 4
Controls: 2; Off-On/Volume, Tuning
See Appendix II (12-15) for the tube diagram for Model 39-70.

Model 39-70B tombstone.

117

Model 39-70F console.

Model 39-71
Frequency Coverage: 530-1720 kc
Power: Battery operated
Tubes Used: 4
Controls: 2; Tuning, Off-On/Volume
See Appendix II (12-16) for the tube diagram for Model 39-71.

Model 39-71T portable receiver.

Model 39-75
Frequency Coverage: 530-1720 kc
Power: Battery operated
Tubes Used: 4
Controls: 2; Off-On/Volume, Tuning
These models feature six pushbuttons for automatic tuning
See Appendix II (12-17) for the tube diagram for Model 39-75.

Model 39-75 was offered in the same cabinets as Model 39-17, shown on page 113.

Model 39-80
Frequency Coverage: 530-1720 kc
Power: Battery operated
Tubes Used: 4
Controls: 2; Off-On/Volume, Tuning
See Appendix II (12-18) for the tube diagram for Model 39-80.

Model 39-80XF console.

Model 39-80B is in a cabinet similar to Model 39-85B tombstone, shown below.

Model 39-85
Frequency Coverage: Two bands.
    Band #1: 540-1720 kc
    Band #2: 5.6-18.0 mc
Power: Battery operated
Tubes Used: 4
Controls: 3; Off-On/Volume, Band Switch, Tuning
These models feature six pushbuttons for automatic tuning
See Appendix II (12-19) for the tube diagram for Model 39-85.

Model 39-85B tombstone.

Model 39-85XF console. *Photo by Edmund DeCann.*

Model 39-116
Frequency Coverage: Three bands.
    Band #1: 540-1720 kc
    Band #2: 1.7-5.8 mc
    Band #3: 5.8-18.0 mc
Power: AC; 115 volts, 50-60 cycles
Tubes Used: 14
Controls: 4; Tone, Off-On/Volume, Tuning, Band Switch
These models feature Mystery (wireless remote) Control
See Appendix II (12-20) for the tube diagram for Model
39-116.

Model 39-116RX console.

Model 39-116PCX Radio-Phonograph.

Model 907

This is a wireless record player which can play records through any radio.

Model 907T table model wireless record player.

Model 907F consolette wireless record player.

Models RP-l, RP-2, RP-3 & RP-4

(all used the same chassis)
Wireless Record Player
Can be adjusted to broadcast between 530-580 kc
Power: AC; 115 volts, 60 cycles
Tubes Used: 2
See Appendix II (12-21) for the tube diagram for Models RP-l, RP-2, RP-3 & RP-4.

Model RP-l wireless record player.

Model RP-2 wireless record player.

Model RP-3 wireless record player.

Model RP-4 consolette wireless record player.

Model TH-l Transitone table model.

Model TH-3 Transitone
Frequency Coverage: 540-1720 kc
Power: AC; 115 volts, 50-60 cycles
Tubes Used: 5
Controls: 2; Off-On/Volume, Tuning

Model TH-3 Transitone table model.

Models TH-4 & TP-4 Transitone
(both used very similar chasses)
Frequency Coverage:
Model TH-4: 540-1720 kc
Model TP-4: Two bands.
    Band #1: 540-1720 kc
    Band #2: 2.3-2.5 mc
Power: 115 volts AC/DC
Tubes Used: 5
Controls: 2; Off-On/Volume, Tuning (plus band switch
in back of TP-4)

Model TH-4 Transitone bakelite table model.

Model TH-l Transitone
Circuit: Tuned Radio Frequency (TRF)
Frequency Coverage: 540-1720 kc
Power: 115 volts AC/DC
Tubes Used: 4; with one ballast
Controls: 2; Off-On/Volume, Tuning

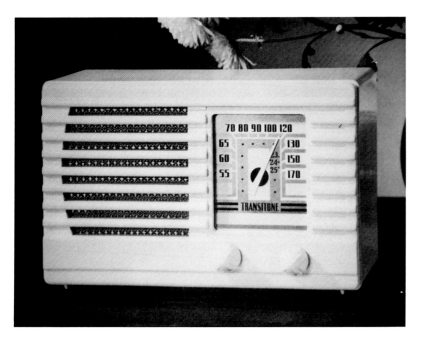

Model TP-4I Transitone bakelite table model with ivory finish.

Model TH-4I Transitone is in a cabinet identical to Model TP-4I Transitone bakelite table model with ivory finish.

Model TP-4 Transitone is in a cabinet identical to Model TH-4 Transitone bakelite table model.

Models TH-5 & TP-5 Transitone
(both used very similar chasses)
Frequency Coverage:
Model TH-5: 540-1720 kc
Model TP-5: Two bands.
    Band #1: 540-1720 kc
    Band #2: 2.3-2.5 mc
Power: 115 volts AC/DC
Tubes Used: 5
Controls: 2; Off-On/Volume, Tuning (plus band switch in back of TP-5)
These models feature six pushbuttons for automatic tuning

Model TH-5 Transitone bakelite table model.

Model TP-5I Transitone bakelite table model with ivory finish.

Model TH-5I Transitone is in a cabinet identical to Model TP-5I Transitone bakelite table model with ivory finish.

Model TP-5 Transitone is in a cabinet identical to Model TH-5 Transitone bakelite table model.

Models TP-10 & TP-11 Transitone
(both used very similar chasses)
Frequency Coverage: Two bands.
    Band #1: 540-1720 kc
    Band #2: 2.3-2.5 mc
Power: 115 volts AC/DC
Tubes Used: 5
Controls: 2; Off-On/Volume, Tuning; plus band switch in back of set Model TP-ll features six pushbuttons for automatic tuning

Model TP-10 Transitone plastic table model.

Model TP-11 Transitone plastic table model.

Model TP-12 Transitone
Frequency Coverage: Two bands.
    Band #1: 540-1720 kc
    Band #2: 2.3-2.5 mc
Power: 115 volts AC/DC
Tubes Used: 5
Controls: 2; Off-On/Volume, Tuning; plus band switch in back of set

Model TP-12 Transitone table model.

Outwardly, the new 1940 Philco models which made their debut in June 1939 were not much different than the 1939 models. But inside, the sets featured such new innovations as locktal tubes and built-in antennas, which allowed much greater flexibility in installation since no outside antenna or ground was required.

While Philco continued to promote its compact table models at the low end of the price scale (fourteen Transitone models were available), the company was also placing greater emphasis on radio-phonograph models at the high end. This year's top of the line model was the 40-516P, a thirteen tube receiver with automatic record changer, three-band coverage, and wireless remote control (by now, the term "Mystery Control" had been dropped). It carried a price tag of $395, and was only one of fourteen different radio-phonographs available from Philco in 1940. In all, there were sixty-three models to choose from in the new Philco line, including a wide variety of consoles, table models, portables and even a chairside model. The RP-l and RP-2 wireless phonographs were still available as well.

Philco's popular wireless remote control was available in four other models in 1940: the 40-205RX, 40-215RX and 40-216RX as well as the 40-510P radio-phonograph.

In January 1940, Philco introduced several new Transitone compact sets which replaced all but one of the earlier models (the TP-lOC was still available). In addition, four new consoles were added to the line including another wireless remote control model. Two new portable models were now available as well.

Along with the radios, phonographs, and radio-phonograph combination models, Philco continued to offer its own line of refrigerators and air conditioners.

By now, the company was in its last months of operation as the Philadelphia Storage Battery Company/Philco Radio & Television Corporation. While Philco's radios would continue to become less distinctive from the competition, sweeping changes were about to take place in the company's structure.

Philco All Year 'Round

Philco had a new slogan for the 1940 season - "Philco All Year 'Round." What exactly did they mean? The fact is, this was only part of the story.

To its dealers, Philco pushed its complete 1940 slogan "Philco All Year 'Round for Profits All Year 'Round." Philco's product line was intended to sell all year, not just during one or two seasons.

Radio sales had always been strongest in the fall and early winter. Manufacturers also made changes to their lineup each January in order to stimulate spring sales. Now, with Philco's air conditioners (made by York Ice Machinery Company) and its new refrigerators, the company's dealers had products that would also sell during the summer, when radio sales were slowest.

By now, most American homes had at least one radio. Because of this, companies were beginning to have a little trouble selling the big, expensive console radios. However, small, compact table models were doing very well as people began to see a need for second or third radios in their homes.

Philco had recognized this trend the year before and introduced its line of compact Transitone table model radios. Small radios were not as profitable as big sets, however, which led some companies to begin offering non-radio items. This is one reason why Philco had begun to sell air conditioners, and then, refrigerators.

It was now possible for the average home to have one or more Philco radios, along with a Cool-Wave air conditioner in the living room or bedroom, and a Philco refrigerator in the kitchen. Of course,

the family car may have had a Philco Transitone car radio in it, as Philco offered radios for several different automobile makes.

As the years passed, Philco continued to add more items to its product line. By the mid-fifties, that same average home could have added a Philco television in the living room, a Philco electric stove in the kitchen, and a Philco freezer in the kitchen or utility room.

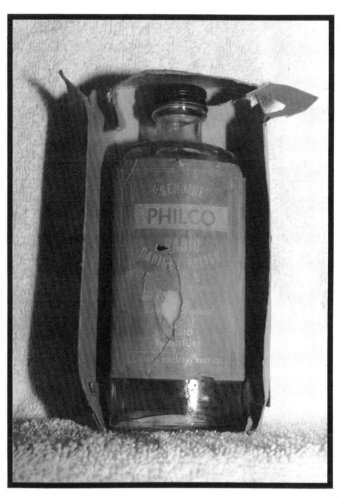

Philco radio cabinet polish. *Photo by John Miller.*

Model 40-74
Frequency Coverage: 530-1600 kc
Power: Battery operated
Tubes Used: 4
Controls: 2; Off-On/Volume, Tuning

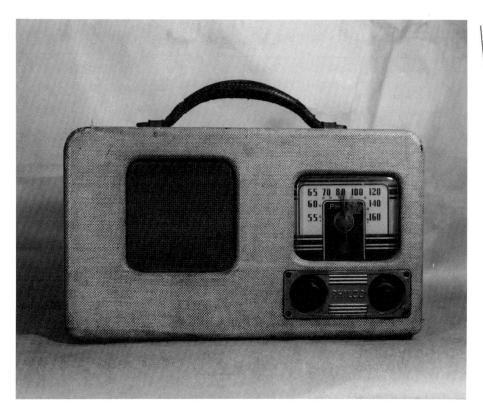

Model 40-74T portable receiver. *Photo by Doug Houston.*

Models 40-81 & 40-82
(both used the same chassis)
Frequency Coverage: 540-1550 kc
Power: Battery operated
Tubes Used: 4
Controls: 2; Off-On/Volume, Tuning

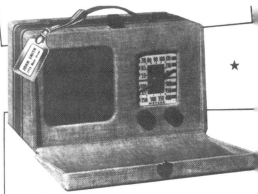

PHILCO 82 T

A smart, brand-new Philco Portable with a protective folding lid that covers the entire front. Lies flat when open. Small, lightweight, self-powered . . . with performance far beyond its size **and price!** Extra long-life battery, super - efficient Loktal Tubes. Built-in Loop Aerial. Attractive case covered in airplane luggage cloth, with handle and tag for name and address.

Model 40-82T portable receiver. *Courtesy of Michael Prosise.*

Model 40-84
Frequency Coverage: 540-1550 kc
Power: 115 volts AC/DC or battery operated
Tubes Used: 5
Controls: 2; Off-On/Volume, Tuning

Model 40-84T is in a portable style cabinet.

Model 40-88
Frequency Coverage: Two bands.
    Band #1: 550-1600 kc
    Band #2: 6.0-18.0 mc
Power: Battery operated
Tubes Used: 5
Controls: 3; Off-On/Volume, Band Switch, Tuning
See Appendix II (13-1) for the tube diagram for Model 40-84.

Model 40-81T portable receiver. *Photo by Doug Houston.*

Model 40-88T portable receiver. *Photo by Doug Houston.*

Model 40-90
Frequency Coverage: 540-1720 kc
Power: Battery operated
Tubes Used: 4
Controls: 2; Off-On/Volume, Tuning

Model 40-90CB was offered in a bakelite table model cabinet.

Model 40-95
Frequency Coverage: 540-1720 kc
Power: Battery operated
Tubes Used: 4
Controls: 2; Off-On/Volume, Tuning

Model 40-95T table model.

Model 40-95F console.

Model 40-100
Frequency Coverage: 540-1720 kc
Power: Battery operated
Tubes Used: 4
Controls: 2; Off-On/Volume, Tuning
These models feature six pushbuttons for automatic tuning

Model 40-100 was offered in table model and console cabinets.

Model 40-105
Frequency Coverage: 540-1720 kc
Power: Battery operated
Tubes Used: 4
Controls: 2; Off-On/Volume, Tuning

Model 40-105 was offered in tombstone and console cabinets.

Model 40-110
Frequency Coverage: Two bands.
    Band #1: 540-1630 kc
    Band #2: 5.4-18.0 mc
Power: Battery operated
Tubes Used: 4
Controls: 3; Off-On/Volume, Tuning, Band Switch
These models feature six pushbuttons for automatic tuning
See Appendix II (13-2) for the tube diagram for Model 40-110.

Model 40-110 was offered in tombstone and console cabinets.

Models 40-115 & 40-124
(both used very similar chasses)
Frequency Coverage: Two bands.
    Band #1: 550-1600 kc
    Band #2: 1.6-3.3 mc
Power: 115 volts AC/DC
Tubes Used: 6
Controls: 3; Off-On/Volume, Band Switch, Tuning
Model 40-124 features six pushbuttons for automatic tuning
See Appendix II (13-3) for the tube diagram for Models 40-115 & 40-124.

Model 40-115C table model. *Photo by Michael Prosise.*

Model 40-124C table model. *Photo by Michael Prosise.*

Models 40-120 & 40-125
(both used very similar chasses)
Frequency Coverage: Two bands.
    Band #1: 540-1600 kc
    Band #2: 1.6-3.3 mc
Power: 115 volts AC/DC
Tubes Used: 6
Controls: 3; Off-On/Volume, Band Switch, Tuning
Model 40-125 features six pushbuttons for automatic tuning
See Appendix II (13-4) for the tube diagram for Models 40-120 & 40-125.

Models 40-130, 40-135, 40-170, 40-503, 40-506, 40-525, 40-526 & 40-527
(all used very similar chasses)
Frequency Coverage: Two bands.
    Band #1: 540-1550 kc
    Band #2: 1.5-3.3 mc
Power: AC; 115 volts, 60 cycles
Tubes Used: 6
Controls: 4; Volume, Off-On/Tone, Tuning, Band Switch
Models 40-135, 40-170, 40-503, 40-506, 40-525, 40-526 & 40-527 feature six pushbuttons for automatic tuning
See Appendix II (13-5) for the tube diagram for Models 40-130, 40-135, 40-170, 40-503, 40-506, 40-525, 40-526 & 40-527.

Model 40-130T table model. *Photo by Doug Houston.*

Model 40-120C table model. *Photo by Jerry McKinney.*

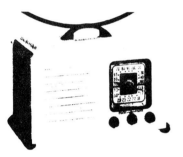

PHILCO 120 C I

Built to receive Television Sound. Super-performance in a compact of gleaming Ivory. Built-in Super Aerial *System* with costly R. F. Stage. Six working, super-efficient Loktal Tubes, Electro-Dynamic Speaker, Automatic Volume Control. AC-DC operation. Gets Standard American, State and City Police, Amateur Stations. Gleaming Ivory finish wood cabinet.

Model 40-120CI table model with ivory finish. *Courtesy of Michael Prosise.*

Model 40-135T table model. *Photo by Doug Houston.*

Model 40-125C table model. *Photo by Doug Houston.*

New Chairside PHILCO

PHILCO 170 CS **Built to receive Television Sound.** Place it beside any chair in your home . . . plug in and play! Built-in Super Aerial *System* eliminates messy aerial and ground wires . . . brings clear tone, **even in noisy locations!** Six Electric Push-Buttons . . . five for favorite stations, one for dial tuning. Concert Grand Speaker, Tone Control, Automatic Volume Control. Gets Standard Broadcasts, State and City Police Calls. Exquisite Walnut chairside cabinet, finished on all sides and mounted on easy rolling casters. All Controls conveniently located on top.

Model 40-170CS chairside set. *Courtesy of Michael Prosise.*

PHILCO 503

Superb tone and performance. Specially designed Audio System and Speaker for fuller, clearer tone on records. **Semi-automatic operation**—just lift lid to play and repeat. "True-tracking." Plays 10-inch and 12-inch records. Built-in Super Aerial System. Six tubes. Gets Standard and Police reception. Built to receive Television Sound.

Model 40-503P table model Radio-Phonograph. *Courtesy of Michael Prosise.*

NEW PHILCO
Radio - Phonograph

MODEL 525 We bring you this brand-new radio-phonograph . . . a triumph in tone, operation and performance. Automatic Record Changer plays 12 ten-inch or 10 twelve-inch records at one loading. Cabinet lid lifts up for access to Phonograph. Radio controls on front of cabinet . . . accessible even with lid closed. Electric Push-Button Tuning, including button for Television Sound reception. Built-in Super Aerial System. No aerial installation . . . just plug in anywhere and play. Lovely Walnut cabinet. See it!

$**100**

TURN THE PAGE...
See all the Other Big-Value Philcos

Model 40-525P Radio-Phonograph. *Courtesy of Michael Prosise.*

Models 40-526P & 40-527P are radio-phonograph consoles.

Models 40-140, 40-145 & 40-507
(all used very similar chasses)
Frequency Coverage: Three bands.
    Band #1: 540-1550 kc
    Band #2: 1.5-3.3 mc
    Band #3: 5.7-18.0 mc
Power: AC; 115 volts, 60 cycles
Tubes Used: 6
Controls: 4; Volume, Off-On/Tone, Tuning, Band Switch
Models 40-145 & 40-507 feature six pushbuttons for automatic tuning
See Appendix II (13-6) for the tube diagram for Models 40-140, 40-145 & 40-507.

Model 40-145T table model. *Photo by Doug Houston.*

AMAZING
Anniversary Special!

PHILCO 506

Model 40-506P Radio-Phonograph console. *Courtesy of Michael Prosise.*

Two different versions of Model 40-140T table model.
Photo at top by Michael Prosise.

Models 40-150, 40-155, 40-180, 40-185 & 40-190 (all used similar chasses)

Frequency Coverage: Three bands.
    Band #1: 540-1550 kc
    Band #2: 1.5-3.4 mc
    Band #3: 6.0-18.0 mc
Power: AC; 115 volts, 60 cycles
Tubes Used: see below
Controls: 4; Off-On/Tone, Volume, Band Switch, Tuning

These models feature eight pushbuttons for automatic tuning

See Appendix II (13-7) for the tube diagram for Models 40-150 & 40-180.

See Appendix II (13-8) for the tube diagram for Models 40-155, 40-185 & 40-190.

PHILCO 185 XX

Built to receive Television Sound. Breaks all records for value at this popular price! Eight tubes. Built-in Super Aerial *System* with **Twin-Loop Aerial. Eight Electric Push-Buttons** for favorite stations. Phonograph and Television Sound. Inclined Sounding Board. Cathedral Speaker. Inclined Control Panel. Highly-figured Walnut cabinet. Foreign and American reception. Police Calls. Ship and Amateurs.

Model 40-185XX console. *Courtesy of Michael Prosise.*

Model 40-190XF console. *Courtesy of Michael Prosise.*

PHILCO 190 XF

Built to receive Television Sound. Rich cabinet beauty coupled with fine radio performance. Eight tubes. Built-in Super Aerial *System,* with Twin-Loop Aerial. **Eight Electric Push-Buttons** for favorite stations. Phonograph and Television Sound. Cathedral Speaker. Impressive cabinet of inlaid Walnut with Inclined Control Panel. American and Foreign reception. Police Calls. Ship and Amateurs.

PHILCO 507

This handsome model brings you the **essentials** of fine tone and luxurious convenience at moderate cost. "Expensive" features, many never offered at this price before. Automatic Record-Changer plays, at one loading, twelve 10-inch or ten 12-inch records . . . **longer by half** than others offer at the same price. Sound output has been **doubled;** the speaker refined for richer tone. You hear recorded music with thrilling richness, depth and life-like reality. American and Foreign receiver with Built-in Super Aerial *System* and six tubes . . . including newly-invented Loktal Tubes. Built to receive Television Sound.

Modle 40-507P Radio-Phonograph console. *Courtesy of Michael Prosise.*

Model 40-150T table model. *Photo by Michael Prosise.*

Model 40-155T table model. *Photo by Edmund DeCann.*

Model 40-180XF console. *Photo by Dennis Osborne.*

Model 40-158
Frequency Coverage: Two bands.
Power: AC; 115 volts, 60 cycles
Tubes Used: 6
Controls: 4; Volume, Off-On, Tuning, Band Switch
See Appendix II (13-9) for the tube diagram for Model 40-158.

Model 40-158F console. *Courtesy of Michael Prosise.*

129

Model 40-160

Frequency Coverage: Two bands.
    Band #1: 540-1550 kc
    Band #2: 1.5-3.35 mc
Power: AC; 115 volts, 60 cycles
Tubes Used: 6
Controls: 4; Volume, Off-On/Tone, Tuning, Band Switch
This model features six pushbuttons for automatic tuning
See Appendix II (13-10) for the tube diagram for Model 40-160.

Model 40-160F is in a cabinet similar to model 40-165F console, shown below.

Model 40-165

Frequency Coverage: Three bands.
    Band #1: 540-1550 kc
    Band #2: 1.5-3.5 mc
    Band #3: 6.0-18.0 mc
Power: AC; 115 volts, 60 cycles
Tubes Used: 6
Controls: 4; Volume, Off-On/Tone, Tuning, Band Switch
This model features six pushbuttons for automatic tuning
See Appendix II (13-11) for the tube diagram for Model 40-165.

Model 40-165F console. *Courtesy of Michael Prosise.*

Model 40-165K was offered in a different console cabinet.

Models 40-195, 40-200 & 40-201

(all used similar chasses)
Frequency Coverage: Three bands.
    Band #1: 540-1550 kc
    Band #2: 1.5-4.0 mc
    Band #3: 6.0-18.0 mc
Power: AC; 115 volts, 60 cycles
Tubes Used: see below
Controls: 4; Tone, Volume, Tuning, Band Switch
These models feature eight pushbuttons for automatic tuning
See Appendix II (13-12) for the tube diagram for Model 40-195 & Model 40-201, Code 122.

See Appendix II (13-13) for the tube diagram for Model 40-200 & Model 40-201, Code 121.

Model 40-195XX console. *Photo by Michael Prosise.*

Get Our Sensational
NEW LOW PRICES!

Model 40-200XX console. *Courtesy of Michael Prosise.*

Model 40-201XX was offered in a console cabinet.

Models 40-205 & 40-510

(both used the same chassis)
Frequency Coverage: 540-1600 kc
Power: AC; 115 volts, 60 cycles
Tubes Used: 12
Controls: 4; Tone, Off-On/Volume, Tuning, Standard/Remote Control
These models feature wireless remote control
See Appendix II (13-14) for the tube diagram for Models 40-205 & 40-510.

Model 40-205RX is in a cabinet identical to Model 40-215RX console, shown on page 131.

Model 40-510P is in a cabinet identical to Model 40-516P Radio-Phonograph console, shown on page 131.

Model 40-215 & 40-217

(both used the same chassis)

Frequency Coverage: Three bands.
    Band #1: 540-1520 kc
    Band #2: 1.4-3.6 mc
    Band #3: 6.0-18.0 mc
Power: AC; 115 volts, 60 cycles
Tubes Used: 12
Controls: 4; Tone, Off-On/Volume, Tuning, Band Switch
These models feature wireless remote control
See Appendix II (13-15) for the tube diagram for Model 40-215 & 40-217.

HILCO 215 RX

Built to receive **Television Sound**. American and Foreign reception . . . **PLUS** Wireless Remote Control. Tunes favorite stations from any room without wires or connections. Built-in Super Aerial *System*, powerful 12-tube circuit and a host of other **quality** features. Handsome Inclined Sounding Board cabinet of figured butt Walnut, with folding lid covering Inclined Control Panel.

HILCO 205 RX

Housed in the same stately cabinet, with the same quality features as the Philco 215RX . . . but covers standard American Broadcasts and State Police Calls only.

Model 40-215RX console. *Courtesy of Michael Prosise.*

Model 40-217RX is in a cabinet similar to model 40-216RX console, shown below.

Models 40-216 & 40-516

(both used the same chassis)

Frequency Coverage: Three bands.
Power: AC; 115 volts, 60 cycles
Tubes Used: 14
Controls: 4; Tone, Off-On/Volume, Tuning, Band Switch
These models feature wireless remote control
See Appendix II (13-16) for the tube diagram for Models 40-216 & 40-516.

Model 40-216RX console. *Courtesy of Michael Prosise.*

Model 40-516P Radio-Phonograph console with wireless remote control unit. *Courtesy of Michael Prosise.*

Models 40-501 & 40-502

(both used very similar chasses)
Frequency Coverage: 540-1700 kc
Power: AC; 115 volts, 60 cycles
Tubes Used: 5
Controls: 3; Tuning, Off-On/Volume, Radio-Phono

Anniversary Specials!

New triumphs in Radio-Phonograph tone, performance and beauty! The circuits of all 1940 Philco Radio-Phonographs are specially designed for richer, clearer, fuller reproduction of phonograph records. Not radio circuits, but **radio-phonograph circuits**, with twice the sound output and large oversize speakers that give you finer tone than you have ever heard from records before!

PHILCO 502

Plays 10-inch and 12-inch records with lid closed . . . reducing needle noise. Tone Control. Five tubes. Gets Standard and State Police reception Built to receive Television Sound.

NO WIRES, OR CONNECT

PHILCO 216 RX

Built to receive Television Sound. The triumphant result of new inventions by Philco engineers. Wireless Remote Control tunes favorite stations from any room without wires or connections. Powerful 14-tube circuit. Built-in Super Aerial *System* with **Twin-Loop** Aerial and costly R. F. Stage. Inclined Sounding Board. Stately inlaid Walnut cabinet, with folding lid to cover Inclined Control Panel. Three tuning ranges cover everything that's interesting at home and abroad.

PHILCO 501

A quality instrument, designed for tone, power and performance far beyond its size . . . and price! Efficient Tone Arm, new type Automatic Switch, featherweight needle touch, improved, true-tracking Crystal Pick-up, self-starting motor, volume control for records and radio. Five tubes. Gets Standard Broadcasts and State Police. Built to receive Television Sound.

Model 40-501P table model Radio-Phonograph. *Courtesy of Michael Prosise.*

Model 40-502P table model Radio-Phonograph. *Courtesy of Michael Prosise.*

Model 40-504

Frequency Coverage: 540-1600 kc
Power: Battery operated
Tubes Used: 4
Controls: 3; Tuning, Off-On/Volume, Radio/Phono

Model 40-504P portable Radio-Phonograph with windup phonograph motor. *Courtesy of Michael Prosise.*

. . and a new PORTABLE RADIO-PHONOGRAPH!

PHILCO 504

Now, a Portable Radio Phonograph, self powered! Enjoy radio and records wherever you go. Plays 10-inch and 12-inch records through the radio circuit. Improved Crystal pick-up insures clear tone. Newly-improved Permanent Field Speaker. Self-contained Loop Aerial. Powerful, even speed, hand-wind motor. Attractive airplane luggage cloth case with space for carrying records.

Models 40-508, 40-509 & 40-515
(all used the same chassis)
Frequency Coverage: Three bands.
  Band #1: 540-1550 kc
  Band #2: 1.5-3.4 mc
  Band #3: 6.0-18.0 mc
Power: AC; 115 volts, 60 cycles
Tubes Used: 8
Controls: 4; Off-On/Tone, Volume, Band Switch, Tuning
These models feature eight pushbuttons for automatic tuning

Model 40-508P Radio-Phonograph console. *Courtesy of Michael Prosise.*

Model 40-509P Radio-Phonograph console. *Courtesy of Michael Prosise.*

Model 40-515P is a Radio-Phonograph console, and was offered in walnut and mahogany cabinets.

Model PT-10
Frequency Coverage: 540-1720 kc
Power: 115 volts AC/DC
Tubes Used: 5
Controls: 2; Tuning, Off-On/Volume

Model PT-10C plastic table model. This model was available in fifty-five different color combinations. *Courtesy of Michael Prosise.*

Models PT-25, PT-27 & PT-39
(all used the same chassis)
Frequency Coverage: 540-1720 kc
Power: 115 volts AC/DC
Tubes Used: 5
Controls: 2; Tuning, Off-On/Volume

Model PT-25 table model. *Photo by Doug Houston.*

Model PT-27 bakelite table model with ivory finish. *Courtesy of Michael Prosise.*

PT-39 Large, beautifully grained Walnut cabinet at amazing low price. Attached Aerial . . . no ground. Covers Standard Broadcasts and State Police.

PT-41 Same with Built-in Loop Aerial. Just plug in and play.

PT-33 *Built-in Loop Aerial* . . . no ground needed. Extremely compact model, with brown bakelite cabinet and molded carrying handle. Covers Standard American Broadcasts and State Police.

Model PT-39 table model. *Courtesy of Michael Prosise.*

Model PT-33 bakelite table model. *Courtesy of Michael Prosise.*

Models PT-26, PT-28 & PT-36
(all used the same chassis)
Frequency Coverage: 540-1580 kc
Power: 115 volts AC/DC
Tubes Used: 5
Controls: 2; Tuning, Off-On/Volume

PT-61 The "Jewel Case." A charming, distinctive cabinet in rare, colorful woods of Mexican Accra, Aspen and Birchwood. *Built-in Loop Aerial.* Covers Standard American Broadcasts and State Police Calls.

Model PT-61 "Jewel Case" table model. *Courtesy of Michael Prosise.*

Model PT-41 is in a cabinet identical to Model PT-39 table model, shown on page 133.

Model PT-35
Frequency Coverage: Two bands.
    Band #1: 540-1720 kc
    Band #2: 2.3-2.5 mc
Power: 115 volts AC/DC
Tubes Used: 5
Controls: 2; Tuning, Off-On/Volume; plus a band switch at back of set

Model PT-26 table model. *Photo by Doug Houston.*

Models PT-28 & PT-36 were offered in table model cabinets.

Models PT-29 & PT-31
(both used the same chassis)
Frequency Coverage: Two bands.
    Band #1: 540-1720 kc
    Band #2: 2.3-2.5 mc
Power: 115 volts AC/DC
Tubes Used: 5
Controls: 2; Tuning, Off-On/Volume; plus a band switch at back of set

Model PT-29 is in a cabinet identical to Model PT-25 table model, shown on page 132.

Model PT-31 is in a cabinet identical to Model PT-27 table model, shown on page 132.

Models PT-33, PT-41 & PT-61
(all used the same chassis)
Frequency Coverage: 540-1580 kc
Power: 115 volts AC/DC
Tubes Used: 5
Controls: 2; Tuning, Off-On/Volume

PT-35 Modern plastic cabinet in gleaming ebony, with dial, knobs and speaker grille in contrasting cream. Attached Aerial . . . no ground needed. Covers Standard American Broadcasts, State and City Police.

Model PT-35 table model. *Courtesy of Michael Prosise.*

Models PT-37, PT-38 & PT-53
(all used the same chassis)
Frequency Coverage: Two bands.
    Band #1: 540-1720 kc
    Band #2: 5.5-19.0 mc
Power: 115 volts AC/DC
Tubes Used: 5
Controls: 2; Tuning, Off-On/Volume; plus a band switch at back of set

Model PT-37 is in a cabinet identical to Model PT-25 table model, shown on page 132.

Models PT-38 & PT-53 were offered in table model cabinets.

Models PT-43 & PT-55
(both used the same chassis)
Frequency Coverage: Two bands
    Band #1: 540-1580 kc
    Band #2: 2.3-2.5 mc
Power: 115 volts AC/DC
Tubes Used: 5
Controls: 2; Tuning, Off-On/Volume; plus a band switch
at back of set

PT-43 *Built-in Loop Aerial.* Rich Walnut cabinet of refreshingly new appearance with handle, dial, knobs and speaker grille of cream plastic. Covers Standard Broadcasts, State and City Police.

Model PT-43 table model. *Courtesy of Michael Prosise.*

PT-55 *Built-in Loop Aerial . . .* plug in anywhere and play. Rich plastic cabinet in ebony, with dial, knobs and speaker grille in contrasting cream with carrying handle. Covers Standard American Broadcasts, State and City Police.

Model PT-55 table model. *Courtesy of Michael Prosise.*

Models PT-45 & PT-47
(both used the same chassis)
Frequency Coverage: 540-1720 kc
Power: 115 volts AC/DC
Tubes Used: 5
Controls: 2; Tuning, Off-On/Volume
These models feature six pushbuttons for automatic
tuning

PT-45 Trim brown bakelite cabinet with Electric Push-Button Tuning and Attached Aerial . . . no ground. Covers Standard American Broadcasts and State Police.

Model PT-45 bakelite table model. *Courtesy of Michael Prosise.*

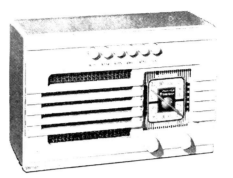

PT-47 A stunning model in a rich ivory bakelite cabinet. Electric Push-Button Tuning. Attached Aerial . . . no ground. Covers Standard American Broadcasts.

PT-51 Same plus State and City Police.

Model PT-47 bakelite table model with ivory finish.
Courtesy of Michael Prosise.

Models PT-46 & PT-48
(both used the same chassis)
Frequency Coverage: 540-1580 kc
Power: 115 volts AC/DC
Tubes Used: 5
Controls: 2; Tuning, Off-On/Volume
These models feature six pushbuttons for automatic
tuning

Models PT-46 & PT-48 were offered in table model
cabinets.

Models PT-49 & PT-51
(both used the same chassis)
Frequency Coverage: Two bands.
    Band #1: 540-1720 kc
    Band #2: 2.3-2.5 mc
Power: 115 volts AC/DC
Tubes Used: 5
Controls: 2; Tuning, Off-On/Volume; plus a band switch
at back of set
These models feature six pushbuttons for automatic
tuning
Model PT-49 was offered in a table model cabinet.

Model PT-51 is in a cabinet identical to Model PT-47
table model, shown above.

Model PT-50
Frequency Coverage: 540-1580 kc
Power: 115 volts AC/DC
Tubes Used: 5
Controls: 2; Tuning, Off-On/Volume

Model PT-50 was offered in a table model cabinet.

Models PT-57 & PT-65
(both used the same chassis)
Frequency Coverage: 540-1580 kc
Power: 115 volts AC/DC
Tubes Used: 5
Controls: 2; Tuning, Off-On/Volume
These models feature six pushbuttons for automatic
tuning

PT-57 A sleek model of brown bakelite with molded handle. Electric Push-Button Tuning. Built-in Loop Aerial. Covers Standard American Broadcasts and State Police.

Model PT-57 bakelite table model. *Courtesy of Michael Prosise.*

PT-65 *Built-in Loop Aerial.* Beautiful hand-rubbed inlaid Walnut cabinet with molded handle. Electric Push-Button Tuning. Covers Standard American Broadcasts.

Model PT-65 table model. *Courtesy of Michael Prosise.*

Model PT-59
Frequency Coverage: Two bands.
   Band #1: 540-1720 kc
   Band #2: 2.3-2.5 mc
Power: 115 volts AC/DC
Tubes Used: 5
Controls: 2; Tuning, Off-On/Volume; plus a band switch at back of set

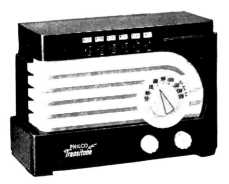

PT-59 Distinctive modern design, brown and amber plastic. Electric Push-Button Tuning. Attached Aerial . . . no ground. Covers Standard American Broadcasts, State and City Police.

Model PT-59 plastic table model. *Courtesy of Michael Prosise.*

Model PT-63
Frequency Coverage: 540-1550 kc
Power: Battery operated
Tubes Used: 4
Controls: 2; Tuning, Off-On/Volume

PT-63 Portable—plays anywhere in or out of doors. Compact, extra light; rich tone; amazing performance. Low-drain, *extra* long battery life. Airplane cloth case with address tag. Standard American Reception.

Model PT-63 portable receiver. *Courtesy of Michael Prosise.*

Model PT-66
Frequency Coverage: 540-1580 kc
Power: 115 volts AC/DC
Tubes Used: 5
Controls: 2; Tuning, Off-On/Volume
This model features six pushbuttons for automatic tuning

Model PT-66 was offered in a table model cabinet.

Model PT-67
Frequency Coverage: Two bands.
   Band #1: 540-1580 kc
   Band #2: 2.3-2.5 mc
Power: 115 volts AC/DC
Tubes Used: 5
Controls: 2; Tuning, Off-On/Volume; plus a band switch at back of set
This model features six pushbuttons for automatic tuning

PT-67 *Built-in Loop Aerial.* Beautiful amber and brown plastic cabinet with molded carrying handle. Electric Push-Button Tuning. Covers Standard American Broadcasts, State and City Police.

Model PT-67 plastic table model. *Courtesy of Michael Prosise.*

Model PT-69
Frequency Coverage: 540-1580 kc
Power: AC; 115 volts, 60 cycles
Tubes Used: 5
Controls: 2; Tuning, Off-On/Volume
This model features an electric clock

Model PT-69 table model. *Photo by Michael Prosise.*

Models TH-14 & TH-16
(both used the same chassis)
Frequency Coverage: 540-1580 kc
Power: 115 volts AC/DC
Tubes Used: 5
Controls: 2; Tuning, Off-On/Volume

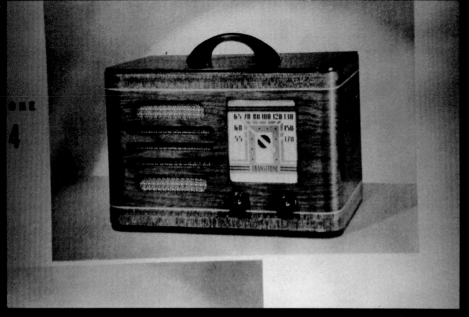

Model TH-14 table model.

Model TH-16 was offered in a table model cabinet.

Models TH-15 & TH-17
(both used the same chassis)
Frequency Coverage: 540-1580 kc
Power: 115 volts AC/DC
Tubes Used: 5
Controls: 2; Tuning, Off-On/Volume
These models feature six pushbuttons for automatic tuning

Model TH-15 table model.

Model TH-17 was offered in a table model cabinet.

Model TH-18
Frequency Coverage: Two bands.
    Band #1: 540-1720 kc
    Band #2: 5.5-19.0 mc
Power: 115 volts AC/DC
Tubes Used: 5
Controls: 3; Tuning, Off-On/Volume, Band Switch

Model TH-18 was available in a table model cabinet.

Model TP-20
Frequency Coverage: Two bands.
    Band #1: 540-1580 kc
    Band #2: 2.3-2.5 mc
Power: 115 volts AC/DC
Tubes Used: 5
Controls: 3; Tuning, Off-On/Volume, Band Switch

Model TP-20 was offered in a table model cabinet.

Model TP-21
Frequency Coverage: 540-1580 kc
Power: 115 volts AC/DC
Tubes Used: 5
Controls: 2; Tuning, Off-On/Volume
This model features six pushbuttons for automatic tuning

Model TP-21 was offered in a table model cabinet.

Philco's new 1941 models continued a trend of looking plainer and more like other manufacturer's sets, in spite of the fact that the company introduced some new cabinet styles for the 1941 season. Seventy-two new models were available, more than Philco had offered at the start of the 1940 season, but fewer radio-phonographs (ten) and fewer Philco radios were offered (thirty-seven), while more Transitone models were in the lineup (twenty).

Philco's newest "first" was unveiled at the start of the new season - the Beam of Light phonograph. It used a pickup unlike any other phonograph had ever used. The large pickup head included a permanent jewel stylus, a small mirror, a lamp, and a photocell. A special coil in the radio, coupled to the set's oscillator tube, produced an RF voltage which lit the lamp. As a record was played, the jewel stylus in the record groove made the mirror vibrate. The lamp shone on the mirror and as the mirror vibrated the photocell picked up the variations of light reflected from the mirror and sent this signal, which was the audio from the record, into the radio. The result was improved tone with less surface noise, and reduced record wear.

A home recording unit was optional with Beam of Light phonographs, enabling the set owner to record as well as play records. (Tape recording did not become a reality until after World War II.)

The 1941 Philco line was headed by the 41-616P. Its chassis was similar to the previous season's 40-516P, but the new version featured a new cabinet, available in walnut or mahogany, with a speaker grille that tilted forward to reveal the phonograph. The Tilt-Front cabinet was a feature of all 1941 Philco console radio-phonographs. Two Tilt-Front models (the 41-616 and 41-611) were designed by Emil I. Harman, while the 41-608 Tilt-Front was designed by Edgar F. Haines.

The new 41-616P also offered four band coverage, wireless remote control, and an automatic record changer with the new Beam of Light pickup. The price remained the same as last season's 40-516P at $395.

Many new Philco sets also offered an improved built-in antenna, along with a new loctal first detector tube, the type XXL.

About the same time the new 1941 models were being introduced, Philco bought controlling interest in the National Union Radio Corporation. National Union's main product was vacuum tubes. This meant that Philco no longer had to rely on outside sources to make its tubes for them.

In July, the Philadelphia Storage Battery Company and the Philco Radio & Television Corporation were merged into a new firm, the Philco Corporation; and for the first time, Philco offered stock to the public.

In November, Philco produced its fifteen millionth radio, which prompted a special promotion, the "Fifteen Millionth Philco Jubilee."

Twenty-two new models were added to the Philco line in early 1941, including three more Transitones, five portables, six radio-phonographs, a table model Beam of Light phonograph, two consoles, four table models and a table model radio with a clock (model 41-KR).

Now let's take a look at some of Philco's 1941 models.

The Beam of Light Phonograph

"Music on a beam of light!" It was now possible to enjoy music with less noise and better tone than conventional record players offered. No, we're not talking about compact discs, but Philco's Beam of Light phonograph pickup. It used a permanent jewel stylus instead of the old steel needle used in many other phonographs since the turn of the century. For years, the steel needles required a heavy weight to track properly. Although the 78 rpm records of the day included abrasive fillers which actually wore the needle down, the heavy tracking forces caused fairly rapid record wear.

Philco promised its new pickup would cut down on record wear. This new pickup, using a light beam reflected off a tiny mirror attached to the stylus, could be considered a forerunner to today's compact disc system. The vibrations of the mirror on the stylus varied, or modulated, the light beam. This was picked up by a photocell and converted to an electrical signal, from which the receiver produced sound. (A complete description of the Beam of Light pickup is given on page *.)

Today's CD players use a laser beam to provide "music on a beam of light" as well. The laser "reads" microscopic holes in the compact disc. The player then converts the modulated beam to a digital signal, and then to sound.

There are differences, of course, between Philco's Beam of Light pickup and today's CD players. Philco's elaborate system was still a mechanical apparatus. It required a stylus riding in the record groove, as conventional phonograph pickups did. A modern CD player has no stylus, as the laser beam does the job without any other apparatus making mechanical contact with the CD's surface.

Philco's Beam of Light was new and different, but it failed to catch on as the new pickup was only used in 1941 and 1942 models. When civilian production resumed late in 1945, Philco no longer offered the Beam of Light. Therefore, this unusual phonograph pickup had a very brief history.

"Tube Saver" tag which was attached to some Philco sets of the period. *Photo by John Miller.*

Models 41-22CL, 41-220 & 41-225
(all used very similar chasses)
Frequency Coverage: Two bands.
    Band #1: 540-1600 kc
    Band #2: 1.6-3.3 mc
Power:
    Model 41-22CL: AC; 115 volts, 60 cycles
    Models 41-220 & 41-225: 115 volts AC/DC
Tubes Used: 6
Controls: 3; Volume, Band Switch, Tuning
Model 41-22CL features an electric clock
Model 41-225 features six pushbuttons for automatic tuning
See Appendix II (14-1) for the tube diagram for Models 41-22CL, 41-220 & 41-225.

Model 41-22CL table model radio with clock. *Photo by Michael Prosise.*

Model 41-220C table model with brown trim. *Photo by Gary Schieffer.*

225 C Six tubes, AC-DC. R.F. Stage. Built-In Super-Aerial System. Electro-Dynamic Speaker. 6 Electric Push-Buttons. Pentode Audio System. Gets Standard Broadcasts, State and City Police, Aircraft, Amateurs.

Model 41-225C table model. *Courtesy of Michael Prosise.*

Model 41-220CI was offered in a table model cabinet with ivory trim.

Model 41-81
Frequency Coverage: 540-1600 kc
Power: Battery operated
Tubes Used: 4
Controls: 2; Tuning, Off-On/Volume

Model 41-81T was offered in a portable case.

Model 41-83
Frequency Coverage: 540-1600 kc
Power: Battery operated
Tubes Used: 5
Controls: 2; Tuning, Off-On/Volume

Model 41-83T was offered in a portable case.

Model 41-84
Frequency Coverage: 540-1600 kc
Power: 115 volts AC/DC or battery operated
Tubes Used: 5
Controls: 2; Tuning, Off-On/Volume

Model 41-84T was offered in a portable case.

Model 41-85
Frequency Coverage: Two bands.
Power: 115 volts AC/DC or battery operated
Tubes Used: 5
Controls: 3; Tuning, Off-On/Volume, Band Switch

Model 41-85T was offered in a portable case.

Model 41-90
Frequency Coverage: 540-1600 kc
Power: Battery operated
Tubes Used: 4
Controls: 2; Tuning, Off-On/Volume

Model 41-90CB is in a cabinet similar to Model 41-230T table model, shown on page 140.

Model 41-95
Frequency Coverage: 540-1600 kc
Power: Battery operated
Tubes Used: 5
Controls: 2; Tuning, Off-On/Volume

Model 41-95T table model.

Model 41-100
Frequency Coverage: 540-1600 kc
Power: Battery operated
Tubes Used: 5
Controls: 2; Tuning, Off-On/Volume
These models feature six pushbuttons for automatic tuning

Model 41-100 was offered in table model and console cabinets.

Model 41-105

Frequency Coverage: Two bands.
Power: Battery operated
Tubes Used: 5
Controls: 3; Tuning, Off-On/Volume, Band Switch
This model features six pushbuttons for automatic tuning

Model 41-105T was offered in a table model cabinet.

Model 41-110

Frequency Coverage: Two bands.
Power: Battery operated
Tubes Used: 6
Controls: 4; Volume, Off-On/Tone, Tuning, Band Switch
See Appendix II (14-2) for the tube diagram for Model
41-110.

Model 41-110K was offered in a console cabinet.

Models 41-221, 41-226 & 41-231

(all used very similar chasses)
Frequency Coverage: Two bands.
   Band #1: 540-1700 kc
   Band #2: 8.8-12.0 mc
Power: 115 volts AC/DC
Tubes Used: 6
Controls: 3; Volume, Band Switch, Tuning
Models 41-226 & 41-231 feature six pushbuttons for
automatic tuning
See Appendix II (14-3) for the tube diagram for Models
41-221, 41-226 & 41-231.

Model 41-226C table model. *Photo by Michael Prosise.*

Model 41-221CI table model with ivory trim.

Model 41-221C table model with brown trim. *Photo by
Michael Prosise.*

Model 41-231T table model. *Photo by Michael Prosise.*

Models 41-230 & 41-235

(both used the same chassis)
Frequency Coverage: Two bands.
   Band #1: 540-1600 kc
   Band #2: 1.6-3.5 mc
Power: AC; 115 volts, 60 cycles
Tubes Used: 7
Controls: 4; Volume, Off-On/Tone, Tuning, Band Switch
See Appendix II (14-4) for the tube diagram for Models
41-230 & 41-235.

Model 41-230T table model. *Photo by Doug Houston.*

235T Seven-tube Superheterodyne including new XXL tubes. Built-in Super Aerial System. Pentode Audio System. Tone Control. Illuminated Dial. Gets Standard Broadcasts, State and City Police, Aircraft, Amateurs

Model 41-235T table model. *Courtesy of Michael Prosise.*

Model 41-240
Frequency Coverage: Two bands.
    Band #1: 540-1700 kc
    Band #2: 9.0-12.0 mc
Power: AC; 115 volts, 60 cycles
Tubes Used: 7
Controls: 4; Volume, Off-On/Tone, Tuning, Band Switch
See Appendix II (14-5) for the tube diagram for Model 41-240.

Model 41-240T table model. *Photo by Doug Houston.*

Models 41-245 & 41-246
(both used very similar chasses)
Frequency Coverage: Three bands.
Power: AC; 115 volts, 60 cycles
Tubes Used: 7
Controls: 4; Volume, Tone, Tuning, Band Switch
These models feature six pushbuttons for automatic tuning
See Appendix II (14-6) for the tube diagram for Models 41-245 & 41-246.

245T Seven tubes. Overseas Wave-Band. New Aerial System. 6 Push-Buttons. Standard, American and Foreign Short-Wave, State, City Police, Aircraft, Amateurs.

246T In same cabinet with American and Foreign Day and Night Short-Wave, State Police, Aircraft, Amateurs.

Model 41-245T table model. *Courtesy of Michael Prosise.*

Model 41-246T is in a cabinet identical to Model 41-245T.

Models 41-250, 41-255 & 41-256
(all used similar chasses)
Frequency Coverage: Three bands.
Power: AC; 115 volts, 60 cycles
Tubes Used: see below
Controls: 4; Bass, Volume, Band Switch, Tuning
These models feature eight pushbuttons for automatic tuning
See Appendix II (14-7) for the tube diagram for Model 41-250.

See Appendix II (14-8) for the tube diagram for Models 41-255 & 41-256.

Model 41-250T table model. *Photo by Doug Houston.*

Model 41-255T table model.

Model 41-256T is in a cabinet identical to Model 41-255T table model.

Model 41-258
Frequency Coverage: Two bands.
Power: 115 volts AC/DC
Tubes Used: 6
Controls: 3; Off-On/Volume, Band Switch, Tuning

Model 41-258F console. *Photo by Michael Prosise.*

Model 41-260
Frequency Coverage: Two bands.
    Band #1: 540-1720 kc
    Band #2: 9.0-12.0 mc
Power: AC; 115 volts, 60 cycles
Tubes Used: 7
Controls: 4; Volume, Tone, Tuning, Band Switch
This model features six pushbuttons for automatic tuning.
See Appendix II (14-9) for the tube diagram for Model 41-260.

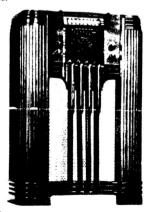

260 F Seven-tube Superheterodyne Circuit, including new XXL tubes. New Overseas Wave-Band. Built-In American and Overseas Aerial System. 6 Electric Push-Buttons. Covers Standard Broadcasts, American and Foreign Short-Wave Programs and State Police Calls.

Model 41-260F console. *Courtesy of Michael Prosise.*

Model 41-265
Frequency Coverage: Three bands.
    Band #1: 540-1720 kc
    Band #2: 2.0-7.0 mc
    Band #3: 9.0-12.0 mc
Power: AC; 115 volts, 60 cycles
Tubes Used: 7
Controls: 4; Volume, Tone, Tuning, Band Switch
This model features six pushbuttons for automatic tuning
See Appendix II (14-10) for the tube diagram for Model 41-265.

Model 41-265K console.

Models 41-280, 41-285, 41-287 & 41-290
(all used similar chasses)
Frequency Coverage: Three bands.
    Band #1: 540-1720 kc
    Band #2: 2.3-7.0 mc
    Band #3: 9.0-12.0 mc
Power: AC; 115 volts, 60 cycles
Tubes Used: see below
Controls: 4; Tone, Volume, Band Switch, Tuning
These models use eight pushbuttons for automatic tuning
See Appendix II (14-11) for the tube diagram for Model 41-280.
See Appendix II (14-12) for the tube diagram for Models 41-285 & 41-287.
See Appendix II (14-13) for the tube diagram for Model 41-290.

Model 41-280X console. *Photo by David Millward.*

285X Nine-tube Superheterodyne Circuit, including new XXL tubes. Overseas Wave-Band. Built-In American and Overseas Aerial System. 8 Electric Push-Buttons. 3 Tuning Bands cover Standard Broadcasts. American and Foreign Short-Wave, State and City Police, Aircraft, Ship, Amateurs.

Model 41-285X console. *Courtesy of Michael Prosise.*

287X Nine-tube Superheterodyne Circuit, including new XXL tubes. New Overseas Wave-Band. American and Overseas Aerial System. 8 Electric Push-Buttons. 3 Tuning Bands cover Standard, American and Foreign Short-Wave, State and City Police, Aircraft, Ship and Amateur broadcasts.

Model 41-287X console. *Courtesy of Michael Prosise.*

290X Ten-tube Superheterodyne Circuit, including new XXL tubes. New Overseas Wave-Band. American and Overseas Aerial System. 8 Electric Push-Buttons. 3 Tuning Bands cover Standard, American and Foreign Short-Wave, State and City Police, Aircraft, Ship and Amateur broadcasts.

Model 41-290X console. *Courtesy of Michael Prosise.*

Models 41-295, 41-300 & 41-315
(all used similar chasses)
Frequency Coverage: Four bands.
Power: AC; 115 volts, 60 cycles
Tubes Used: see below
Controls: 4; Tone, Volume, Tuning, Band Switch
These models feature eight pushbuttons for automatic tuning
See Appendix II (14-14) for the tube diagram for Model 41-295.

See Appendix II (14-15) for the tube diagram for Models 41-300 & 41-315.

295X Eleven-tube Superheterodyne Circuit, with new XXL tubes. Overseas Wave-Band. Built-In American and Overseas Aerial System. Cathedral Speaker. Variable Tone Control. 8 Push-Buttons: 7 for favorite stations, one for "On-Off." Highly-figured Walnut cabinet. Illuminated Shifting Band Arrow. 4 Tuning Bands cover American, American and Foreign Short-Wave, State and City Police, Aircraft, Ship and Amateurs.

296X Nine-tube Superheterodyne in same cabinet. 3 Tuning Bands.

Model 41-295X console. *Courtesy of Michael Prosise.*

Model 41-300X console.

Model 41-315X is in a cabinet identical to Model 41-316RX console, shown on page 142.

Model 41-296
Frequency Coverage: Three bands.
   Band #1: 540-1720 kc
   Band #2: 2.3-7.0 mc
   Band #3: 9.0-12.0 mc
Power: AC; 115 volts, 60 cycles
Tubes Used: 9
Controls: 4; Tone, Volume, Tuning, Band Switch
This model features eight pushbuttons for automatic tuning
See Appendix II (14-16) for the tube diagram for Model 41-296.

Model 41-296X is in a cabinet identical to Model 41-295X console, shown above.

Model 41-316
Frequency Coverage: Four bands.
   Band #1: 540-1720 kc
   Band #2: 2.3-7.0 mc
   Band #3: 9.0-12.0 mc
   Band #4: 13.5-18.0 mc
Power: AC; 115 volts, 60 cycles

Tubes Used: 15
Controls: 4; Tone, Off-On/Volume, Tuning, Band Switch
This model features wireless remote control and eight pushbuttons
for automatic tuning
See Appendix II (14-17) for the tube diagram for Model 41-316.

Separate remote unit uses a 30 tube.

316RX Wireless Remote Control tunes favorite programs from any room without wires or connections. 15-tube Superheterodyne Circuit, with new XXL tubes. Overseas Wave-Band. Variable Tone Control. 8 Push-Buttons: 7 for favorite stations, 1 "Remote." Walnut Cabinet. 4 Tuning bands cover American, American and Foreign Short-Wave, State and City Police, Aircraft, Ship and Amateurs.

315X Twelve-tube Superheterodyne in same cabinet, without Remote Control.

Model 41-316RX console with wireless remote control unit. *Courtesy of Michael Prosise.*

Model 41-601
Frequency Coverage: 540-1600 kc
Power: AC; 115 volts, 60 cycles
Tubes Used: 5
Controls: 2; Off-On/Volume, Tuning
See Appendix II (14-18) for the tube diagram for Model 41-601.

601P 5-tube Phonograph Circuit. New light weight Crystal Pickup: flexible, long-life needle plays 1000 records. Tone Control. Gets Standard Programs, State Police.

Model 41-601P table model Radio-Phonograph. *Courtesy of Michael Prosise.*

Model 41-602
Frequency Coverage: 540-1600 kc
Power: AC; 115 volts, 60 cycles
Tubes Used: 5
Controls: 2; Off-On/Volume, Tuning
See Appendix II (14-19) for the tube diagram for Model 41-602.

Model 41-602P is a radio-phonograph set.

Model 41-603
Frequency Coverage: Two bands.
Power: AC; 115 volts, 60 cycles
Tubes Used: 6
Controls: 3; Off-On/Volume, Band Switch, Tuning
See Appendix II (14-20) for the tube diagram for Model 41-603.

Model 41-603P table model Radio-Phonograph. *Photo by Doug Houston.*

Models 41-604, 41-605 & 41-607
(all used the same chassis)
Frequency Coverage: Two bands.
Power: AC; 115 volts, 60 cycles
Tubes Used: 6
Controls: 3; Off-On/Volume, Band Switch, Tuning
See Appendix II (14-21) for the tube diagram for Models 41-604, 41-605 & 41-607.

Model 41-604P is a radio-phonograph set.

Model 41-607 is a radio-phonograph console.

Models 41-608 & 41-609
(both used the same chassis)
Frequency Coverage: Two bands.
Power: AC; 115 volts, 60 cycles
Tubes Used: 9
Controls: 4; Volume, Tone, Tuning, Band Switch
These models feature six pushbuttons for automatic tuning and the Beam of Light phono pickup
See Appendix II (14-22) for the tube diagram for Models 41-608 & 41-609.

608P Philco Photo-Electric Reproducer. 9-tube Phonograph Circuit. Automatic Record Changer. Tilt-Front Cabinet. New Overseas Wave-Band. Built-In American and Overseas Aerial System. 6 Electric Push-Buttons. Covers Standard Broadcasts, American and Foreign Short-Wave, State Police Calls.

Model 41-608P Radio-Phonograph console. *Courtesy of Michael Prosise.*

609P Philco Photo-Electric Reproducer. 9-tube Phonograph Circuit. Automatic Record Changer. Tilt-Front Cabinet, authentic Hepplewhite design. New Overseas Wave-Band. Built-In American and Overseas Aerial System. 6 Electric Push-Buttons. Covers Standard Broadcasts, American and Foreign Short-Wave, State Police Calls.

Model 41-609P Radio-Phonograph console. *Courtesy of Michael Prosise.*

Models 41-610 & 41-611
(both used the same chassis)
Frequency Coverage: Three bands.
Power: AC; 115 volts, 60 cycles
Tubes Used: 10
Controls: 4; Tone, Volume, Band Switch, Tuning
These models feature eight pushbuttons for automatic tuning and the Beam of Light phono pickup
See Appendix II (14-23) for the tube diagram for Models 41-610 & 41-611.

Model 41-603P table model Radio-Phonograph. *Photo by Michael Prosise.*

610P Philco Photo-Electric Reproducer. 10-tube Phonograph Circuit. Automatic Record Changer. Tilt-Front Cabinet. New Overseas Wave-Band. Built-In American and Overseas Aerial System. 8 Electric Push-Buttons. 3 Tuning Bands cover Standard Programs. American and Foreign Short-Wave, State and City Police. Aircraft, Ship, Amateurs.

Model 41-610P Radio-Phonograph console. *Courtesy of Michael Prosise.*

611P Philco Photo-Electric Reproducer. 10-tube Phonograph Circuit. Deluxe Inter-Mix Record Changer. Tilt-Front cabinet, authentic Chippendale design. New Overseas Wave-Band. Built-In American and Overseas Aerial System. 8 Electric Push-Buttons. 3 Tuning Bands cover Standard Broadcasts, American and Foreign Short-Wave, State and City Police. Aircraft, Ship and Amateurs.

Model 41-611P Radio-Phonograph console. *Courtesy of Michael Prosise.*

Model 41-616
Frequency Coverage: Four bands.
    Band #1: 540-1720 kc
    Band #2: 2.3-7.0 mc
    Band #3: 9.0-12.0 mc
    Band #4: 13.5-18.0 mc
Power: AC; 115 volts, 60 cycles
Tubes Used: 16
Controls: 4; Tone, Off-On/Volume, Tuning, Band Switch
This model features eight pushbuttons for automatic tuning, wireless remote control, and the Beam of Light phono pickup
See Appendix II (14-24) for the tube diagram for Model 41-616.

Separate remote unit uses a 30 tube.

616P Wireless Remote Control. Photo-Electric Reproducer. 16 tubes. Inter-Mix Record Changer. Tilt-Front Georgian cabinet. Overseas Wave-Band. Built-In American and Overseas Aerial System. 8 Electric Push-Buttons. 4 Tuning Bands. Standard, American and Foreign Short-Wave, State, City Police, Aircraft, Ship, Amateurs.

Model 41-616P Radio-Phonograph console with wireless remote control unit. This cabinet was available in walnut or mahogany. *Courtesy of Michael Prosise.*

Model 41-620 Phonograph

Power: AC; 115 volts, 60 cycles
Tubes Used: 6
Controls: 2; Tone, Volume
This model features the Beam of Light phono pickup
See Appendix II (14-25) for the tube diagram for Model 41-620 phonograph.

Model 41-620P is a table model phonograph.

Models 41-623, 41-624 & 41-625

(all used the same chassis)
Frequency Coverage: Two bands.
Power: AC; 115 volts, 60 cycles
Tubes Used: 7
Controls: 3; Off-On/Volume, Band Switch, Tuning
These models feature the Beam of Light phono pickup
See Appendix II (14-26) for the tube diagram for Models 41-623, 41-624 & 41-625.

623 P Philco Photo Electric Reproducer. 7-tube Phonograph Circuit. Overseas Wave Band. Gets Standard Broadcasts, American and Foreign Short-Wave, State Police.

Model 41-623P table model Radio-Phonograph. *Courtesy of Michael Prosise.*

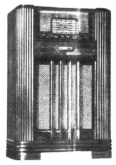

624 P Philco Photo Electric Reproducer. 7-tube Phonograph Circuit. Tilt-Front Cabinet. Overseas Wave-Band. Built-In American and Overseas Aerial System. Permanent Magnet Speaker. Self-starting Phonograph Motor. Covers Standard Broadcasts, American and Foreign Short-Wave, State Police.

Model 41-624P Radio-Phonograph console. *Courtesy of Michael Prosise.*

625 P Philco Photo-Electric Reproducer. 7-tube Phonograph Circuit. Automatic Record Changer. Tilt-Front Cabinet. Beam Power Audio System. Permanent Magnet Speaker. Overseas Wave-Band. Built-In American and Overseas Aerial System. Covers Standard Broadcasts, American and Foreign Short-Wave, State Police Calls.

Model 41-625P Radio-Phonograph console. *Courtesy of Michael Prosise.*

Model 41-629

Frequency Coverage: Two bands.
Power: AC; 115 volts, 60 cycles
Tubes Used: 9
Controls: 4; Tone, Volume, Band Switch, Tuning
This model features six pushbuttons for automatic tuning and the Beam of Light phono pickup
See Appendix II (14-27) for the tube diagram for Model 41-629.

629 P Philco Photo-Electric Reproducer. 9-tube Phonograph Circuit. Automatic Record-Changer. Tilt-Front Cabinet. New Overseas Wave-Band. Built-In American and Overseas Aerial System. Six Electric Push-Buttons. Covers Standard Broadcasts, American and Foreign Short-Wave, State Police Calls.

Model 41-629P Radio-Phonograph console. *Courtesy of Michael Prosise.*

Model 41-695

Frequency Coverage: 540-1720 kc
Power: Battery operated
Tubes Used: 5
Controls: 4; Off-On/Volume, Tone, Radio-Phono, Tuning

Model 41-695P is a radio-phonograph set with a windup phonograph motor.

Model 41-714

Frequency Coverage: Three bands.
Power: AC; 115/230 volts, 50-60 cycles
Tubes Used: 6
Controls: 4; Volume, Off-On/Tone, Tuning, Band Switch
See Appendix II (14-28) for the tube diagram for Model 41-714.

Model 41-714T was offered in a table model cabinet.

Model 41-722

Frequency Coverage: Three bands.
　　Band #1: 540-1720 kc
　　Band #2: 2.3-7.1 mc
　　Band #3: 7.0-22.0 mc
Power: AC; 115/230 volts, 50-60 cycles
Tubes Used: 6
Controls: 4; Tone, Off-On/Volume, Band Switch, Tuning
See Appendix II (14-29) for the tube diagram for Model 41-722.

Model 41-722T was offered in a table model cabinet.

Model 41-758

Frequency Coverage: Three bands.
　　Band #1: 540-1720 kc
　　Band #2: 2.3-7.1 mc
　　Band #3: 7.0-22.0 mc
Power: AC; 115/230 volts, 50-60 cycles
Tubes Used: 8
Controls: 4; Off-On/Tone, Volume, Band Switch, Tuning
See Appendix II (14-30) for the tube diagram for Model 41-758.

Model 41-758T was offered in a table model cabinet.

Model 41-788

Frequency Coverage: Eight bands.
　　Band #1: 540-1720 kc
　　Band #2: 2.3-7.2 mc
　　Band #3: 7.2-22.0 mc
　　Band #4: 9.4-9.9 mc
　　Band #5: 11.4-12.0 mc
　　Band #6: 14.8-15.6 mc
　　Band #7: 17.3-18.2 mc
　　Band #8: 20.9-21.9 mc
Power: AC; 115/230 volts, 50-60 cycles
Tubes Used: 11
Controls: 5; Off-On/Tone, Volume, Normal Tuning, Band Switch, Spread Tuning
See Appendix II (14-31) for the tube diagram for Model 41-788.

Model 41-788T table model. *Photo by Michael Prosise.*

Model 41-841

Frequency Coverage: 540-1600 kc
Power: 115 volts AC/DC or battery operated
Tubes Used: 5
Controls: 2; Off-On/Volume, Tuning

Model 41-841T portable receiver.

Models 41-842, 41-843 & 41-844

(all used the same chassis)
Frequency Coverage: 540-1600 kc
Power: 115 volts AC/DC or battery operated
Tubes Used: 7
Controls: 2; Off-On/Volume, Tuning

842T New, 7-tube Portable. AC, DC or Battery. Tremendous power, amazing sensitivity, selectivity. Beaver graining case.

Model 41-842T portable receiver. *Courtesy of Michael Prosise.*

843T New, 7-tube Portable. AC, DC or Battery. New, fine tone, extreme sensitivity, selectivity. Walrus grain case with drop cover.

Model 41-843T portable receiver. *Courtesy of Michael Prosise.*

844T New, 7-tube Portable. AC, DC or Battery. Amazing performance, superb tone. Solid Walnut case with Beaver graining.

Model 41-844T portable receiver. *Courtesy of Michael Prosise.*

Model 41-851
Frequency Coverage: Two bands.
Power: 115 volts AC/DC or battery operated
Tubes Used: 5
Controls: 3; Off-On/Volume, Band Switch, Tuning

851T AC, DC or Battery Portable. American and Foreign *Short-Wave*. Standard Broadcasts. 5 tubes. Ostrich graining case.

Model 41-851T portable receiver. *Courtesy of Michael Prosise.*

Model 41-KR
Frequency Coverage: 540-1600 kc
Power: AC; 115 volts, 60 cycles
Tubes Used: 5
Controls: 2; Off-On/Volume, Tuning

Model 41-KR is a table model radio with a built-in clock.

Models 41-RP1, 41-RP2 & 41-RP6
(all used very similar chasses)
Wireless Record Player
Can be adjusted to broadcast between 530-580 kc
Power: AC; 115 volts, 60 cycles
Tubes: 2

Model 41-RP1 wireless record player. *Photo by Doug Houston.*

Model 41-RP2 wireless record player. *Photo by Doug Houston.*

RP6 Wireless Record Player in table-type cabinet. Top space for radio or decorations; easy-sliding drawer contains Record Player. Plays records through your radio circuit without wires or connections to radio. Crystal Pick up. Plays 10" and 12" records.

Model 41-RP6 wireless record player. *Courtesy of Michael Prosise.*

Models PT-2 & PT-6
(both used the same chassis)
Frequency Coverage: 540-1600 kc
Power: 115 volts AC/DC
Tubes Used: 5
Controls: 2; Off-On/Volume, Tuning
See Appendix II (14-32) for the tube diagram for Models
PT-2 & PT-6.

Model PT-2 table model. *Photo by Bob Schafbuch.*

Model PT-6 was offered in a table model cabinet.

Model PT-12
Frequency Coverage: 540-1600 kc
Power: 115 volts AC/DC
Tubes Used: 5
Controls: 2; Off-On/Volume, Tuning
See Appendix II (14-33) for the tube diagram for Model
PT-12.

Model PT-12 was offered in a table model cabinet.

Models PT-30 & PT-49
(both used the same chassis)
Frequency Coverage: 540-1600 kc
Power: 115 volts AC/DC
Tubes Used: 5
Controls: 2; Off-On/Volume, Tuning
See Appendix II (14-34) for the tube diagram for Models
PT-30 & PT-49.

Model PT-30 table model. *Photo by Jerry McKinney.*

Model PT-49 was offered in a table model cabinet.

Models PT-42 & PT-44
(both used the same chassis)
Frequency Coverage: 540-1600 kc
Power: 115 volts AC/DC
Tubes Used: 5
Controls: 2; Off-On/Volume, Tuning

See Appendix II (14-35) for the tube diagram for Models
PT-42 & PT-44.

Model PT-44 table model.

Model PT-42 was offered in a table model cabinet.

Model PT-87
Frequency Coverage: 540-1600 kc
Power: 115 volts AC/DC or battery operated
Tubes Used: 5
Controls: 2; Off-On/Volume, Tuning
See Appendix II (14-36) for the tube diagram for Model
PT-87.

87 PT AC, DC or Battery
Portable. 5 tubes, Built-In
Loop Aerial, Permanent
Magnet Speaker. Smart,
lightweight case.

Model PT-87 portable receiver. *Courtesy of Michael Prosise.*

Model PT-89
Frequency Coverage: 540-1600 kc
Power: Battery operated
Tubes Used: 4
Controls: 2; Off-On/Volume, Tuning

Model PT-89C portable receiver. *Photo by Doug Houston.*

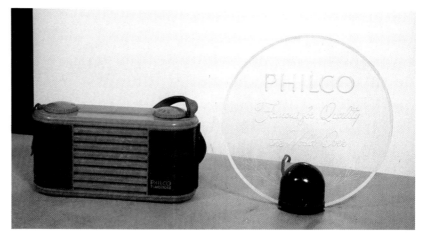

In June 1941, Philco unveiled its new 1942 model line. Many models and cabinets carried on virtually unchanged from the 1941 season. Changes that were made included more pushbuttons on some models that not only selected favorite standard broadcast stations, but also switched the set to the desired band and turned the set off.

Many new Philco sets also featured a frequency modulation (FM) band. FM offered greater fidelity with very little static. However, Philco was comparatively late in offering FM as many other manufacturers had begun to use the new technology a year or two earlier (including Stromberg-Carlson, which had introduced one of the first radios to include FM reception in 1940).

Another "16" headed up the 1942 line. The 42-1016 radio-phonograph console was similar to the 1941 model 41-616 in that it was housed in the same cabinet (available only in walnut this year) and featured four tuning bands, wireless remote control, and the Beam of Light phono pickup. New features included an FM band and twelve pushbuttons - six for selection of favorite standard broadcast stations, four for selecting one of its tuning ranges, one to select phonograph operation and one for wireless remote control.

Another new feature offered in the 42-1016, as well as most of Philco's other radio-phonograph consoles, was a variable speed automatic record changer with a built-in stroboscope, which the set owner could use to see if the phonograph turntable was revolving at the correct speed of 78 rpm or not. A slower speed was also provided, which was not compatible with early 33-1/3 rpm records, since the Philco phonograph's slow speed was 39 rpm!

Sixty-one new 1942 Philco radios, phonographs, and radio-phonograph combinations were available, along with Philco's refrigerators and air conditioners. As America prepared for possible entry into World War II, Philco also began to produce military items for the government. Meanwhile, in September, Philco's experimental television station W3XE in Philadelphia became commercial station WPTZ.

There would be none of the traditional mid-season changes in Philco's line this season. On December 7, Japan attacked Pearl Harbor. The next day, America declared war. In early 1942, civilian production of such items as radios and refrigerators ended as Philco joined other American manufacturers in converting to war work for the government.

Before Philco's civilian production ended, however, the company put two car radios on the market that had been converted for home use. One of the sets (model A-801) was in a chairside cabinet.

There would be no more new radios for the civilian market until late 1945.

Model 42-22CL
Frequency Coverage: Two bands.
Band #1: 540-1600 kc
Band #2: 1.6-3.3 mc
Power: AC; 115 volts, 60 cycles
Tubes Used: 6
Controls: 3; Off-On/Volume, Band Switch, Tuning
This model features an electric clock

Model 42-22CL table model radio with clock.

Model 42-121
Frequency Coverage: 540-1720 kc
Power: Battery operated
Tubes Used: 4
Controls: 2; Off-On/Volume, Tuning
See Appendix II (15-1) for the tube diagram for Model 42-121.

Model 42-121CB table model.

Model 42-122
Frequency Coverage: 540-1720 kc
Power: Battery operated
Tubes Used: 5
Controls: 2; Off-On/Volume, Tuning
See Appendix II (15-2) for the tube diagram for Model 42-122.

Model 42-122T table model.

Model 42-123
Frequency Coverage: 540-1720 kc
Power: Battery operated
Tubes Used: 5
Controls: 2; Off-On/Volume, Tuning
See Appendix II (15-3) for the tube diagram for Model
42-123.

Model 42-123F console.

Models 42-124 & 42-125
(both used the same chassis)
Frequency Coverage: Two bands.
    Band #1: 540-1720 kc
    Band #2: 5.7-15.5 mc
Power: Battery operated
Tubes Used: 5
Controls: 4; Tone, Off-On/Volume, Tuning, Band Switch
See Appendix II (15-4) for the tube diagram for Models
42-124 & 42-125.

Model 42-125K console.

Model 42-126
Frequency Coverage: Two bands.
    Band #1: 540-1720 kc
    Band #2: 5.7-15.5 mc
Power: Battery operated
Tubes Used: 6
Controls: 4; Tone, Off-On/Volume, Tuning, Band Switch
This model features six pushbuttons for automatic tuning
See Appendix II (15-5) for the tube diagram for Model
42-126.

Model 42-124T table model.

Model 42-126T table model.

Models 42-321 & 42-PT10
(both used the same chassis)
Frequency Coverage: 540-1600 kc
Power: 115 volts AC/DC
Tubes Used: 6
Controls: 2; Off-On/Volume, Tuning
See Appendix II (15-6) for the tube diagram for Models
42-321 & 42-PT10.

Model 42-PT10 bakelite table model.

Model 42-322
Frequency Coverage: Two bands.
    Band #1: 540-1720 kc
    Band #2: 8.7-15.5 mc
Power: 115 volts AC/DC
Tubes Used: 6
Controls: 3; Off-On/Volume, Tuning, Band Switch
See Appendix II (15-7) for the tube diagram for Model
42-322.

Model 42-321T table model. *Photo by Doug Houston.*

Model 42-321TI table model with ivory trim.

Model 42-322T table model. *Photo by Michael Prosise.*

Model 42-323
Frequency Coverage: Two bands.
    Band #1: 540-1720 kc
    Band #2: 9.0-15.5 mc
Power: 115 volts AC/DC
Tubes Used: 6
Controls: 3; Off-On/Volume, Tuning, Band Switch
See Appendix II (15-8) for the tube diagram for Model
42-323.

Model 42-323T table model.

Model 42-327
Frequency Coverage: Two bands.
    Band #1: 540-1720 kc
    Band #2: 9.0-15.5 mc
Power: 115 volts AC/DC
Tubes Used: 6
Controls: 3; Volume, Tuning, Band Switch
This model features six pushbuttons for automatic tuning
See Appendix II (15-9) for the tube diagram for Model 42-327.

Model 42-335T table model.

Model 42-340
Frequency Coverage: Two bands.
    Band #1: 540-1720 kc
    Band #2: 9.0-15.0 mc
Power: AC; 115 volts, 60 cycles
Tubes Used: 7
Controls: 4; Off-On/Tone, Volume, Band Switch, Tuning
See Appendix II (15-11) for the tube diagram for Model 42-340.

Model 42-327T table model.

Model 42-335
Frequency Coverage: Two bands.
    Band #1: 540-1600 kc
    Band #2: 1.6-3.3 mc
Power: AC; 115 volts, 60 cycles
Tubes Used: 7
Controls: 4; Off-On/Tone, Volume, Band Switch, Tuning
See Appendix II (15-10) for the tube diagram for Model 42-335.

Model 42-340T table model.

Model 42-345
Frequency Coverage: Three bands.
    Band #1: 540-1720 kc
    Band #2: 2.3-7.0 mc
    Band #3: 9.0-15.5 mc
Power: AC; 115 volts, 60 cycles
Tubes Used: 7
Controls: 4; Tone, Volume, Band Switch, Tuning
This model features six pushbuttons for automatic tuning
See Appendix II (15-12) for the tube diagram for Model 42-345.

Model 42-345T table model. *Photo by Doug Houston.*

Model 42-355T table model. *Photo by John Okolowicz.*

Model 42-350
Frequency Coverage: Three bands.
    Band #1: 540-1720 kc
    Band #2: 9.0-15.0 mc
    Band #3: 42-50 mc (FM)
Power: AC; 115 volts, 60 cycles
Tubes Used: 7
Controls: 4; Tone, Volume, Band Switch, Tuning
This model features six pushbuttons for automatic tuning; also Frequency Modulation (FM) reception

Model 42-350T table model. *Photo by Doug Houston.*

Models 42-355 & 42-390
(both used the same chassis)
Frequency Coverage: Three bands.
    Band #1: 540-1720 kc
    Band #2: 9.0-15.0 mc
    Band #3: 42-50 mc (FM)
Power: AC; 115 volts, 60 cycles
Tubes Used: 8
Controls: 4; Volume, Bass, Treble, Tuning
This model features nine pushbuttons for automatic tuning, band selection, and on-off; also Frequency Modulation (FM) reception
See Appendix II (15-13) for the tube diagram for Models 42-355 & 42-390.

Model 42-390X console.

Model 42-358
Frequency Coverage: 540-1620 kc
Power: 115 volts AC/DC
Tubes Used: 6
Controls: 2; Off-On/Volume, Tuning
See Appendix II (15-14) for the tube diagram for Model 42-358.

Model 42-358F console.

Model 42-360

Frequency Coverage: Two bands.
    Band #1: 540-1720 kc
    Band #2: 9.0-15.5 mc
Power: AC; 115 volts, 60 cycles
Tubes Used: 7
Controls: 4; Volume, Off-On/Tone, Tuning, Band Switch
See Appendix II (15-15) for the tube diagram for Model
42-360.

Model 42-365

Frequency Coverage: Three bands.
    Band #1: 540-1720 kc
    Band #2: 2.3-6.7 mc
    Band #3: 9.0-15.5 mc
Power: AC; 115 volts, 60 cycles
Tubes Used: 7
Controls: 4; Volume, Tone, Tuning, Band Switch
This model features six pushbuttons for automatic tuning
See Appendix II (15-16) for the tube diagram for Model
42-365.

Model 42-365K console.

Model 42-380

Frequency Coverage: Three bands.
    Band #1: 540-1720 kc
    Band #2: 2.3-7.0 mc
    Band #3: 9.0-15.5 mc
Power: AC; 115 volts, 60 cycles
Tubes Used: 8
Controls: 4; Volume, Bass, Treble, Tuning
This model feature nine pushbuttons for automatic tuning, band selection, and on-off selection
See Appendix II (15-17) for the tube diagram for Model
42-380.

Model 42-360F console. *Photo by Michael Prosise.*

Model 42-380X console.

Model 42-395
Frequency Coverage: Three bands.
    Band #1: 540-1720 kc
    Band #2: 9.0-15.5 mc
    Band #3: 42-50 mc (FM)
Power: AC; 115 volts, 60 cycles
Tubes Used: 9
Controls: 4; Volume, Bass, Treble, Tuning
This model features ten pushbuttons for automatic tun-
ing, band selection, and on-off selection; also Frequency
Modulation (FM) reception
See Appendix II (15-18) for the tube diagram for Model
42-395.

Band #3: 14.4-18.0 mc
    Band #4: 42-50 mc (FM)
Power: AC; 115 volts, 60 cycles
Tubes Used: 11
Controls: 4; Volume, Bass, Treble, Tuning
This model features ten pushbuttons for automatic tun-
ing, band selection, and on-off selection; also Frequency
Modulation (FM) reception
See Appendix II (15-19) for the tube diagram for Model
42-400.

Model 42-395X console.

Model 42-400
Frequency Coverage: Four bands.
    Band #1: 540-1720 kc
    Band #2: 9.0-12.0 mc

Closed (above) and open (right) views of Model 42-400X
console. *Photos by Wayne King.*

Model 42-620 Phonograph
Power: AC; 115 volts, 60 cycles
Tubes Used: 6
Controls: 2; Tone, Volume
This model features the Beam of Light phono pickup

Model 42-620P table
model phonograph.

Models 42-842, 42-843 & 42-844
(all used the same chassis)
Frequency Coverage: 540-1600 kc
Power: 115 volts AC/DC or battery operated
Tubes Used: 7
Controls: 2; Off-On/Volume, Tuning

Model 42-854T portable receiver.

Model 42-842T portable receiver. *Photo by Doug Houston.*

Model 42-843T is in a cabinet similar to Model 42-853T portable receiver, shown below.

Model 42-844T is in a cabinet similar to Model 42-854T portable receiver, shown below.

Models 42-853 & 42-854
(both used the same chassis)
Frequency Coverage: Two bands.
    Band #1: 540-1600 kc
    Band #2: 5.7-15.5 mc
Power: 115 volts AC/DC or battery operated
Tubes Used: 7
Controls: 3; Off-On/Volume, Tuning, Band Switch
See Appendix II (15-20) for the tube diagram for Models 42-853 & 42-854.

Courtesy of Michael Prosise.

Model 42-1001
Frequency Coverage: 540-1600 kc
Power: AC; 115 volts, 60 cycles
Tubes Used: 5
Controls: 2; Off-On/Volume, Tuning
See Appendix II (15-21) for the tube diagram for Model 42-1001.

Model 42-853T portable receiver.

Model 42-1001 table model Radio-Phonograph. *Photo by Michael Prosise.*

Model 42-1002
Frequency Coverage: 540-1600 kc
Power: AC; 115 volts, 60 cycles
Tubes Used: 6
Controls: 2; Off-On/Volume, Tuning
See Appendix II (15-22) for the tube diagram for Model 42-1002.

Model 42-1002 table model Radio-Phonograph.

Model 42-1003
Frequency Coverage: Two bands.
    Band #1: 540-1720 kc
    Band #2: 9.0-15.5 mc
Power: AC; 115 volts, 60 cycles
Tubes Used: 7
Controls: 3; Off-On/Volume, Band Switch, Tuning
This model features the Beam of Light phono pickup
See Appendix II (15-23) for the tube diagram for Model 42-1003.

Model 42-1003 table model Radio-Phonograph. *Photo by Doug Houston.*

Model 42-1004
Frequency Coverage: 540-1600 kc
Power: AC; 115 volts, 60 cycles
Tubes Used: 6
Controls: 2; Off-On/Volume, Tuning
See Appendix II (15-24) for the tube diagram for Model 42-1004.

Model 42-1004 Radio-Phonograph console.

Model 42-1005
Frequency Coverage: Two bands.
    Band #1: 540-1720 kc
    Band #2: 9.0-15.5 mc

Power: AC; 115 volts, 60 cycles
Tubes Used: 7
Controls: 3; Off-On/Volume, Band Switch, Tuning
This model features the Beam of Light phono pickup
See Appendix II (15-25) for the tube diagram for Model 42-1005.

Model 42-1005 Radio-Phonograph console.

Model 42-1006

Frequency Coverage: 540-1600 kc
Power: AC; 115 volts, 60 cycles
Tubes Used: 7
Controls: 3; Off-On/Volume, Radio-Phono, Tuning
See Appendix II (15-26) for the tube diagram for Model 42-1006.

Model 42-1006 Radio-Phonograph console.

Models 42-1008 & 42-1009

(both used the same chassis)
Frequency Coverage: Two bands.
   Band #1: 540-1720 kc
   Band #2: 9.0-15.5 mc
Power: AC; 115 volts, 60 cycles
Tubes Used: 9
Controls: 4; Tone, Volume, Band Switch, Tuning
These models feature six pushbuttons for automatic tuning and the Beam of Light phono pickup
See Appendix II (15-27) for the tube diagram for Models 42-1008 & 42-1009.

Model 42-1008 Radio-Phonograph console. *Photo by Doug Houston.*

Model 42-1009 Radio-Phonograph console. This cabinet was available in walnut or mahogany.

Models 42-1010 & 42-1011
(both used the same chassis)
Frequency Coverage: Three bands.
Power: AC; 115 volts, 60 cycles
Tubes Used: 10
Controls: 4; Volume, Bass, Treble, Tuning
These models feature ten pushbuttons for automatic
tuning, band selection, radio-phono selection, and on/
off; also the Beam of Light phono pickup
See Appendix II (15-28) for the tube diagram for Models
42-1010 & 42-1011.

These models feature ten pushbuttons for automatic
tuning, band selection, radio-phono selection, and on/
off; also Frequency Modulation (FM) reception and the
Beam of Light phono pickup
See Appendix II (15-29) for the tube diagram for Models
42-1012 & 42-1013.

Model 42-1010 Radio-Phonograph console.

Model 42-1012 Radio-Phonograph console.

Model 42-1011 Radio-Phonograph console. This cabinet was available in walnut or mahogany.

Models 42-1012 & 42-1013
(both used the same chassis)
Frequency Coverage: Three bands.
    Band #1: 540-1720 kc
    Band #2: 9.0-15.5 mc
    Band #3: 42-50 mc (FM)
Power: AC; 115 volts, 60 cycles
Tubes Used: 10
Controls: 4; Volume, Bass, Treble, Tuning

Closed (left) and open (right) views of Model 42-1013
Radio-Phonograph console in walnut. This model was
also available in a mahogany cabinet.

Model 42-1015

Frequency Coverage: Four bands.
    Band #1: 540-1720 kc
    Band #2: 9.0-12.0 mc
    Band #3: 13.3-18.0 mc
    Band #4: 42-50 mc (FM)
Power: AC; 115 volts, 60 cycles
Tubes Used: 12
Controls: 4; Volume, Bass, Treble, Tuning
This model features twelve pushbuttons for automatic tuning, band selection, radio-phono selection, and on-off selection; also Frequency Modulation (FM) reception and the Beam of Light phono pickup
See Appendix II (15-30) for the tube diagram for Model 42-1015.

Model 42-1015 Radio-Phonograph console. This cabinet was available in walnut or mahogany.

Model 42-1016

Frequency Coverage: Four bands.
    Band #1: 540-1720 kc
    Band #2: 9.0-12.0 mc
    Band #3: 14.4-18.0 mc
    Band #4: 42-50 mc (FM)
Power: AC; 115 volts, 60 cycles
Tubes Used: 16
Controls: 4; Volume, Bass, Treble, Tuning
This model features twelve pushbuttons for automatic tuning, band selection, radio-phono selection, and on-off selection; also Frequency Modulation (FM) reception, the Beam of Light phono pickup, and wireless remote control
See Appendix II (15-31) for the tube diagram for Model 42-1016.

Model 42-1016 Radio-Phonograph console.

Models 42-KR3 & 42-KR5

(both used the same chassis)
Frequency Coverage: 540-1600 kc
Power:
    Model 42-KR3: 115 volts AC/DC
    Model 42-KR5: AC; 115 volts, 60 cycles
Tubes Used: 5
Controls: 2; Off-On/Volume, Tuning
Model 42-KR5 features an electric clock
See Appendix II (15-32) for the tube diagram for Models 42-KR3 & 42-KR5.

Models 42-KR3 and 42-KR5 were available in table model cabinets.

Models 42-PT2, 42-PT4 & 42-PT7

(all used the same chassis)
Frequency Coverage: 540-1600 kc
Power: 115 volts AC/DC
Tubes Used: 5
Controls: 2; Off-On/Volume, Tuning
See Appendix II (15-33) for the tube diagram for Models 42-PT2, 42-PT4 & 42-PT7.

Model 42-PT4 bakelite table model with ivory finish.

Model 42-PT7 table model. *Photo by Michael Prosise.*

Model 42-PT2 is in a cabinet identical to Model 42-PT10 table model, shown on page 149.

Models 42-PT25 & 42-PT26
Frequency Coverage: 540-1720 kc
Power: 115 volts AC/DC
Tubes Used: 5
Controls: 2; Off-On/Volume, Tuning

Model 42-PT87 portable receiver. *Photo by Doug Houston.*

Models 42-PT91, 42-PT92, 42-PT93, 42-PT94 & 42-PT95
Frequency Coverage: 540-1600 kc
Power: 115 volts AC/DC
Tubes Used: 5
Controls: 2; Off-On/Volume, Tuning

Model 42-PT25 plastic table model.

Model 42-PT26 is in a cabinet identical to Model 42-PT25 table model.

Models 42-PT87 & 42-PT88
Frequency Coverage: 540-1600 kc
Power: 115 volts AC/DC or battery operated
Tubes Used: 5
Controls: 2; Off-On/Volume, Tuning

Model 42-PT91 bakelite table model. *Photo by Michael Prosise.*

Model 42-PT88 portable receiver.

Model 42-PT92 bakelite table model with ivory finish, sitting on its original box. *Photo by Wayne King.*

Model 42-PT95 table model with ivory trim. *Photo by Gary Schieffer.*

Model A-801
Frequency Coverage: 540-1600 kc
Power: AC; 115 volts, 60 cycles
Tubes Used: 8
Controls: 3; Off-On/Volume, Tone, Tuning
See Appendix II (15-34) for the tube diagram for Model A-801.

Model 42-PT93 table model. *Photo by Doug Houston.*

Model A-801 chairside set.

Model 42-PT94 table model. *Photo by Gary Schieffer.*

During World War II, Philco not only produced radio and electrical equipment for tanks and aircraft, but also items such as fuses for artillery shells and aerial bombs, and the "Bazooka" rocket projectile. The company also operated a school for training military personnel to install, operate and maintain airborne electronic equipment.

Philco received 21 Army-Navy "E" flags during the war, which were awarded on the basis of efficiency in meeting production quotas.

During the war, problems developed with Philco's 1942 model automatic record changers. Their main problem was control of the turntable speed. Several of these sets would develop an erratic control of the speed. During the war, replacement parts for radios were difficult (and sometimes impossible) to obtain, so after the war ended, Philco ultimately came out with a kit to change the 1942 two speed (39 rpm/78 rpm) changers to 78 rpm single speed operation for improved performance.

Following victory in 1945, Philco went back to its normal business of manufacturing radios, radio-phonographs, refrigerators and air conditioners. The company introduced its new 1946 line in late 1945, which now included electric freezers as well.

However, one important post-war item was missing from the Philco line - television. Philco had conducted research into television since 1928, when the company first became involved in radio manufacturing, and had operated its own TV station since 1932. But now, while Philco was pushing refrigerators and freezers, RCA was wasting no time in introducing its famous 630TS television receiver.

Philco's own line of TV sets was not introduced until the summer of 1947. The new 1948 models included the 48-1000, which was housed in an unusual table model cabinet in which the portion of the cabinet which covered the ten inch picture tube protruded from the rest of the enclosure. Meanwhile, RCA had quickly become the leader in television, while another Philco competitor, Zenith (which had also waited until the 1948 season to introduce a line of television receivers) soon became number two, leaving Philco in third place.

Philco had been in the storage battery business since 1906. The company sold its storage battery division in 1948. By 1949, electric ranges had been added to the diverse Philco product line.

The fifties would see Philco begin research into transistors, which had been invented in 1948. In the mid fifties, Philco began to experiment with another new technology -computers. Meanwhile, trouble was brewing as the company was reaching out in too many different directions.

In the summer of 1958, Philco's new 1959 product line included some of the most unusual TV sets ever made. The Predicta series featured a twenty-one inch picture tube inside a plastic enclosure and mounted on a swivel base. Most were mounted on top of a cabinet which housed the television circuitry. The 1959 Predicta models included table models (designed by Richard J. Whipple and Severin Jonassen) and a console (designed by Catherine S. Winkler and Richard Whipple). One model (the Predicta Tandem) was reminiscent of Philco's chairside radios with separate speakers from the early and mid thirties as it featured a separate picture tube, connected to the control cabinet by a twenty-five foot cable. Predicta televisions were quite popular initially.

One year later, Philco introduced yet another "first" - a fully transistorized, battery operated, portable TV. The Safari used a two inch picture tube, but had a built-in magnifier to make the picture look larger. It was lightweight, and could also operate on standard AC household current. It was designed by Emil Harman, who had also designed some of Philco's elaborate radio-phonograph consoles in the early forties.

Meanwhile, problems had developed with the Predicta sets. The picture tubes were burning out or going weak prematurely. Smaller seventeen inch picture tubes were used in all but one 1960 Predicta

model. Better replacement picture tubes were now available for the 1959 models, but the damage had been done as the 1960 models did not sell as well as the earlier models had. To make matters worse, Philco's Predicta models were difficult to service.

1961 found Philco in serious trouble, as the company lost over four million dollars in the first half of the year. However, the company was saved from complete ruin as the Ford Motor Company purchased Philco late in the year. Ford soon returned the Philco-Ford Division to profitability.

By the early seventies, a great deal of interest had developed in antiques and collectibles of all types. At the time, there was a small number of antique radio collectors, but many people - collectors and non-collectors alike - were rediscovering the classic old cathedral radios in attics, second hand stores, and relatively new places like "flea markets." Philco-Ford responded to this trend by marketing a transistorized, scaled down, AM-FM replica of the famous Philco model 90 cathedral.

In 1974, Ford sold its Philco division to General Telephone & Electronics, which also owned Sylvania. Then in 1981, Sylvania and Philco were acquired from GTE by North American Philips Corporation. The Philco name, after enjoying a resurgence of popularity in the sixties, had fallen into decline and nearly disappeared. But now, under the ownership of Philips Consumer Electronics Company (a division of North American Philips), Philco is once again a major producer of television receivers.

Philco Radio Receivers Sold in Canada, Showing U.S. Equivalents (where available and/or applicable)

Model	Type of Cabinet	U.S. Equiv.	Power	Tubes Used
1928				
521	Table Model	511	AC	7
522	Table Model	512	AC	7
523	Table Model	513	AC	7
524	Table Model	514	AC	7
525	Table Model	515	AC	7
541	Lowboy	531	AC	7
561	Highboy	551	AC	7
1929				
82	Console	86	AC	8
82	Lowboy	86	AC	8
82	Highboy	86	AC	8
62	Table Model	65	AC	6
62	Lowboy	65	AC	6
62	Highboy	65	AC	6
62	Deluxe highboy	65	AC	6
83	Lowboy	87	AC	8
83	Highboy	87	AC	8
83	Deluxe highboy	87	AC	8
92	Table Model	95	AC	9
92	Lowboy	95	AC	9
92	Highboy	95	AC	9
92	Deluxe highboy	95	AC	9
1930				
73	Table Model	76	AC	7
73	Console	76	AC	7
73	Lowboy	76	AC	7

Model	Type	Chassis	Power	Tubes
73	Highboy	76	AC	7
73	Deluxe highboy	76	AC	7
20-A	Cathedral	20	AC	7
20-A	Consolette	20	AC	7
30	Lowboy	30	Battery	8
30	Highboy	30	Battery	8
41	Lowboy	41	DC	6
41	Highboy	41	DC	6
77-A	Table Model	77	AC	7
77-A	Console	77	AC	7
77-A	Lowboy	77	AC	7
96-A	Table Model	96	AC	9
96-A	Lowboy	96	AC	9
96-A	Highboy	96	AC	9
296-A	Radio-Phono	296	AC	9
296-A	Concert Grand		AC	9

1931

Model	Type	Chassis	Power	Tubes
35	Cathedral	35	Battery	7
35	Highboy	35	Battery	7
50-A	Cathedral	50	AC	5
50-A	Lowboy	50	AC	5
70-A	Cathedral	70	AC	7
70-A	Highboy	70	AC	7
270-A	Radio-Phono	270	AC	7
370-A	Chairside	370	AC	7
90-A	Cathedral	90	AC	9
90-A	Lowboy	90	AC	9
90-A	Highboy	90	AC	9
111-A	Lowboy	111	AC	11
111-A	Highboy	111	AC	11
211-A	Radio-Phono	211	AC	11
112-A	Lowboy	112	AC	11
112-A	Highboy	112	AC	11
212-A	Radio-Phono	212	AC	11
220-A	Radio-Phono	220	AC	7

1932

Model	Type	Chassis	Power	Tubes
4-A	SW Converter	4	AC	3
43-AH	Highboy	43H	AC	8
51-AB	Cathedral	51	AC	5
51-AL	Lowboy	51L	AC	5
551-A	Colonial Clock	551	AC	5
52-AL	Lowboy	52L	AC	5
71-AB	Cathedral	71B	AC	7
71-AL	Lowboy	71L	AC	7
71-AH	Highboy	71H	AC	7
570-A	Grandfather Clock	570	AC	7
22-AL	Radio-Phono	22L	AC	7
91-AH	Highboy	91H	AC	9
91-AX	Console	91X	AC	9
112-AX	Console	112X	AC	11

1933

Model	Type	Chassis	Power	Tubes
80-AB	Cathedral	80B	AC	4
316-AL	Lowboy	16L	AC	11
318-AH	Highboy	18H	AC	8
319-AL	Lowboy	19L	AC	6
319-AX	Console		AC	6
337-L	Lowboy	37L	Battery	5
357-AC	Table Model	57C	AC	4
360-AB	Cathedral	60B	AC	5
360-AL	Lowboy	60L	AC	5
389-AB	Cathedral	89B	AC	6
391-AL	Lowboy	91L	AC	9
391-AX	Console	91X	AC	9

1934

Model	Type	Chassis	Power	Tubes
316-AX	Console	16X	AC	11
318-AL	Lowboy	18L	AC	8
318-AMX	Console		AC	8
338-B	Cathedral	38B	Battery	5
338-L	Lowboy	38L	Battery	5
344-AB	Cathedral	44B	AC	6
344-AL	Lowboy		AC	6
358-AC	Table Model	58C	AC	4
360-AMB	Tombstone	60MB	AC	5
360-AL	Lowboy	60L	AC	5
360-AH	Highboy		AC	5

1935*

Model	Type	Chassis	Power	Tubes
316-AB	Tombstone	16B	AC	11
316-AL	Lowboy	16L	AC	11
316-AX	Console	16X	AC	11
334-B	Cathedral	34B	Battery	7
334-L	Lowboy	34L	Battery	7
338-B	Cathedral	38B	Battery	5
338-L	Lowboy	38L	Battery	5
345-AC	Table Model	45C	AC	6
345-AL	Lowboy	45L	AC	6
359-AS	Table Model	59S	AC	4
366-AB	Cathedral	66B	AC	5
366-AL	Lowboy	66L	AC	5
3118-AB	Cathedral	118B	AC	8
3118-AL	Lowboy		AC	8
3118-AH	Highboy	I18H	AC	8
3118-AX	Console	118X	AC	8

1936

Model	Type	Chassis	Power	Tubes
338-B	Cathedral	38B	Battery	5
338-F	Console	38F	Battery	5
359-AC	Table Model	59C	AC	4
384-AB	Cathedral	84B	AC	4
3116-AB	Tombstone	116B	AC	11
3116-AX	Console	116X	AC	11
3610-AB	Tombstone	610B	AC	5
3610-AF	Console	610F	AC	5
3610-AL	Console		AC	5
3610-AT	Table Model	610T	AC	5
3623-B	Tombstone	623B	Battery	6
3623-F	Console	623F	Battery	6
3630-AB	Tombstone	630B	AC	6
3630-AX	Console	630X	AC	6
3635-AK	Console		AC	6
3650-AB	Tombstone	650B	AC	8
3650-AX	Console	650X	AC	8
3655-AK	Console		AC	8
3665-AX	Console	665X	AC	10

1937

Model	Type	Chassis	Power	Tubes
37-333B	Cathedral	37-33B	Battery	5
37-333F	Console	37-33F	Battery	5
37-338B	Tombstone	37-38B	Battery	6
37-338J	Console	37-38J	Battery	6
37-361AB	Cathedral	37-61B	AC	5
37-361AF	Console	37-61F	AC	5

Model	Type	Model 2	Power	Tubes
37-384AB	Cathedral	37-84B	AC	4
37-3116AX	Console	37-116X	AC	15
37-3600AC	Table Model	37-600C	AC	4
37-3610AT	Table Model	37-610T	AC	5
37-3610AJ	Console	37-610J	AC	5
37-3623B	Tombstone	37-623B	Battery	6
37-3623J	Console	37-623J	Battery	6
37-3624B	Tombstone	37-624B	6 Volt	6
37-3624X	Console		6 Volt	6
37-3630AB	Tombstone		AC	6
37-3630AX	Console	37-630X	AC	6
37-3640AB	Tombstone	37-640B	AC	7
37-3640AX	Console	37-640X	AC	7
37-3650AX	Console	37-650X	AC	8
37-3670AX	Console	37-670X	AC	11
37-361ABN	Tombstone	37-61B	AC	5
37-310AX	Console	37-10X	AC	9
37-311AX	Console	37-11X	AC	10

1938

Model	Type	Model 2	Power	Tubes
38-C2AXX	Console	38-2XX	AC	11
38-C3AXX	Console	38-3XX	AC	9
38-C4AXX	Console	38-4XX	AC	8
38-C7AT	Table Model	38-7T	AC	6
38-C7ACS	Chairside	38-7CS	AC	6
38-C7AXX	Console	38-7XX	AC	6
38-C9AT	Table Model	38-9T	AC	6
38-C9AX	Console		AC	6
38-C10AT	Table Model	38-10T	AC	5
38-C10AF	Console	38-10F	AC	5
38-C10AK	Console		AC	5
38-C12AT	Table Model	38-12T	AC	5
38-C116AXX	Console	38-116XX	AC	15
38-C324T	Table Model		Battery	4
38-C325T	Table Model		Battery	6
38-C325K	Console		Battery	6
38-C623T	Table Model	38-623T	Battery	6
38-C623X	Console		Battery	6
38-C624T	Table Model	38-624T	6 Volt	6
38-C624X	Console		6 Volt	6

1939

Model	Type	Model 2	Power	Tubes
3A1ACB	Table Model			5
3A1ACBI	Table Model			5
3A2AT	Table Model			5
3A5ACB	Table Model			5
3A5ACBI	Table Model			5
3A5AT	Table Model			5
3B4CB	Table Model		Battery	4
3B4F	Console		Battery	4
317AT	Table Model	39-17T	AC	5
317AF	Console	39-17F	AC	5
319AT	Table Model	39-19T	AC	5
319AF	Console	39-19F	AC	5
330AT	Table Model	39-30T	AC	6
330AXF	Console		AC	6
331AXF	Console	39-31XF	AC	6
340AXX	Console	39-40XX	AC	8
341AXX	Console		AC	8
371T	Portable	39-71T	Battery	4
3116ARX	Console	39-116RX	AC	14
C324T	Table Model		Battery	4
C623T	Table Model		Battery	6
C623X	Console		Battery	6
C624X	Console		6 Volt	6
TP34	Table Model	TP-4	AC/DC	5

Model	Type	Model 2	Power	Tubes
TP34I	Table Model	TP-4I	AC/DC	5
372T	Portable		Battery	4

1940

Model	Type	Model 2	Power	Tubes
10T	Table Model	PT-25	AC/DC	5
10TI	Table Model	PT-27	AC/DC	5
12T	Table Model		AC/DC	5
14T	Table Model		AC/DC	5
15AT	Table Model	40-140T	AC	6
16AT	Table Model	40-145T	AC	6
17AT	Table Model		AC	7
18AF	Console	40-165F	AC	6
19AXF	Console	40-180XF	AC	7
21AXX	Console	40-195XX	AC	10
22AXF	Console		AC	8
23AK	Console		AC	6
24T	Table Model		AC/DC	5
25T	Table Model		AC/DC	6
26T	Table Model		AC/DC	5
27T	Table Model		AC/DC	5
201T	Table Model		Battery	4
202F	Console		Battery	4
203T	Table Model		Battery	5
203F	Console		Battery	5
204T	Table Model		Battery	4
205F	Console		Battery	4
206T	Table Model		Battery	4
330AXF	Console		AC	6
331AXF	Console		AC	6
340AXX	Console		AC	8
341AXX	Console		AC	8
3116ARX	Console		AC	14
401T	Portable	40-81T	Battery	4
402T	Portable	40-88T	Battery	5
403T	Portable		Battery	4
404P	Port. w/phono	40-504P	Battery	4
701	Radio-Phono		AC	7
28T	Table Model		AC/DC	5
29T	Table Model		AC/DC	5
31T	Table Model		AC/DC	5
31TI	Table Model		AC/DC	5

1941

Model	Type	Model 2	Power	Tubes
32AT	Table Model		AC	5
33AT	Table Model		AC	6
34AT	Table Model		AC	7
35AT	Table Model		AC	5
36T	Table Model		AC/DC	5
36TI	Table Model		AC/DC	5
37AT	Table Model		AC	6
38AT	Table Model		AC	8
39AT	Table Model		AC	6
40AK	Console		AC	6
41AK	Console		AC	7
42AX	Console		AC	8
43AX	Console		AC	10
44AX	Console		AC	11
45T	Table Model		AC/DC	5
45TI	Table Model		AC/DC	5
207T	Table Model		Battery	4
209T	Table Model		Battery	5
210T	Table Model		Battery	4
211T	Table Model		Battery	5
212F	Console		Battery	5
213K	Console		Battery	5
214T	Table Model		Battery	6

301RP	Wireless phono 41-RPl		AC	2
405T	Portable		Battery	4
406T	Portable		Battery	5
407T	Portable		AC/DC/Batt.	5
408T	Portable		AC/DC/Batt.	5
702AP	Table Radio-Phono		AC	5
703AP	Radio-Phono		AC	5
715P	Radio-Phono		AC	9
35ATD	Table Model		AC	5
41AJ	Console		AC	7
46T	Table Model		AC/DC	5
46TI	Table Model		AC/DC	5
46TL	Table Model		AC/DC	5
46TLI	Table Model		AC/DC	5
47AT	Table Model		AC	5

1942

33AT	Table Model		AC	6
34AT	Table Model		AC	7
37AT	Table Model		AC	6
44AXD	Console		AC	11
45T	Table Model	42-PT25	AC/DC	5
45TI	Table Model		AC/DC	5
48AT	Table Model		AC	5
49AT	Table Model		AC	5
50AT	Table Model		AC	6
51AT	Table Model		AC	7
52AF	Console		AC	5
53AX	Console		AC	6
54AX	Console		AC	8
55AT	Table Model		A	8
56AX	Console		AC	8
702AP	Table Radio-Phono		AC	5
703AP	Radio-Phono		AC	5
705P**	Radio-Phono		AC	9
715AP	Radio-Phono		AC	9

* From 1935 on Philco models are listed in this chart by model year
  instead of calendar year.
** 60 cycle model. All other Canadian Philco sets are designed to
  operate on 25 to 40 cycle alternating current.
Source: Philco Corporation of Canada Limited, *List of Philco Sets
Sold in Canada to Date,* July 1, 1942.

Glossary

AC tube - A tube with a filament designed to operate on alternating current.

AF - Abbreviation for audio frequency.

Bakelite block condenser - A condenser (also called a *capacitor*), sealed inside a small black bakelite container. Used exclusively by Philco, these bakelite blocks sometimes contain two condensers or even a condenser and a resistor.

Battery eliminator - A device operated from normal household current, which supplies some or all of the voltages required to operate a battery-operated radio receiver, eliminating the need for batteries.

Beam power tube - A special tube which is somewhat similar to a pentode tube except that it has special electrodes between the screen grid and the plate instead of a suppressor grid. These *beam-confining* electrodes concentrate the flow of electrons into a beam, improving the tube's performance. *See also* **pentode** *and* **vacuum tube**.

Cold cathode rectifier - A tube filled with a special gas and usually having a cathode and one or two plates. These tubes do not have a filament or heater (*see* **vacuum tube**). Examples of cold cathode rectifiers include the Raytheon types BA, BH and BR, and type 0Z4.

Field coil - Several turns of insulated wire wound around an iron core called a "pot." This coil, when energized by direct current, creates a magnetic field which is necessary for the electro-dynamic speaker it is a part of to operate.

Full-wave rectifier - A tube with a cathode and two plates. Used to convert alternating current to direct current.

Half-wave rectifier - A tube with a cathode and one plate. Used to convert alternating current to direct current.

High Fidelity - In the thirties, high fidelity was defined as the reproduction of audio frequencies from 50 to 7,500 cycles. Today, 20 to 20,000 cycles (now called *hertz*) is considered to be high fidelity reproduction.

"LOC" terminal post - A terminal post on Philco radios made between 1928 and 1930. The "LOC" is an abbreviation for *local*. When a jumper wire was connected between the Philco receiver's LOC and antenna terminal posts, a condenser connected between the LOC post and the 110 volt AC line allowed the house wiring to act as an antenna. It was called LOC (local) since this arrangement did not work as well as a normal outdoor antenna, but would pick up *local* stations acceptably.

Locktal tube - A vacuum tube somewhat similar to an octal tube, except that it had smaller pins and a large locking pin in the middle which would lock the tube in place in its socket. Also spelled *loctal* and *loktal*. *See also* **octal tube**.

Neutrodyne - A radio receiver similar to a TRF set, but designed with small *neutralizing* condensers in the RF stages which could be adjusted to minimize or *neutralize* oscillation. This circuit was invented by Professor L.A. Hazeltine of the Hazeltine Laboratories. *See also* **TRF**.

Non-regenerative - A radio circuit that did not oscillate (squeal).

Pentode - A vacuum tube with five elements (cathode, control grid, screen grid, suppressor grid and plate). *See also* **vacuum tube**.

Push-pull output - An audio output circuit used in more expensive radio receivers in which two tubes are used to amplify both halves of each audio cycle. This circuit supplies more power to the speaker than a single audio output tube would.

Regenerative receiver - A radio receiver which has a detector tube in which the signal is fed back through the tube many times to increase amplification of the signal. Its major drawback is that if too much signal is fed back, the tube will oscillate (squeal). The circuit was invented by Edwin H. Armstrong.

RF - Radio frequency.

Screen grid tube - A vacuum tube with four elements (cathode, control grid, screen grid and plate). Also called *tetrode*. *See also* **vacuum tube**.

Shielding - Metal covering over an electronic component (such as a coil or transformer). It is done to prevent unwanted pickup or re-radiation of signals.

Superheterodyne - A radio receiver in which a received signal is mixed with another signal produced by the set's local oscillator to produce a third signal, known as the intermediate frequency (IF). IF amplifiers can then amplify the signal much more efficiently than a TRF or Neutrodyne set can. The signal is then detected (converted to an audio signal), amplified further, and sent to the speaker. Also invented by Edwin H. Armstrong.

TRF - Tuned radio frequency. This circuit, along with its closely related cousin, the Neutrodyne, was very popular during the twenties. TRF sets receive a signal and amplify it at the same frequency, unlike the superheterodyne which converts the signal to a lower (intermediate) frequency. The result was that these types of sets, which usually covered the standard AM broadcast band, were required to amplify signals between 550 and 1500 kc. Tuned circuits covering this broad range do not operate as efficently as do super-heterodyne IF stages, which only have to amplify one frequency (the IF), which is why the superheterodyne ultimately became the circuit used in all radios and televisions.

Triode - A vacuum tube with three elements (cathode, control grid and plate). Some Philco receivers have audio output pentode tubes connected as triodes, in which the plate and screen grid are tied together, as are the suppressor grid and the cathode (at the tube socket). *See also* **vacuum tube**.

Tuned circuit - A coil and variable condenser connected together, which can be made to resonate (tuned to) a desired frequency.

Variable-mu screen grid - Also called *super control screen grid* and *remote-cutoff tetrode*. This tube is constructed much like an ordinary screen grid tube except for its control grid, which is spaced wider at its center than at the ends. With this design, the tube's amplification does not vary directly in proportion to the control grid voltage.

Vacuum tube - An electronic device which, depending on its design and the circuit it is used in, can amplify or rectify a signal. The most basic vacuum tube contains two elements (a cathode and a plate). Some cathodes are directly heated filaments, while other cathodes are indirectly heated by a *heater*. All vacuum tubes are evacuated of air, then sealed.

Note: When writing to any of the addresses below, enclose a self-addressed stamped envelope for a reply.

Antique Radio Publications

Antique Radio Classified, P.O. Box 2, Carlisle, MA 01741
The Horn Speaker, P.O. Box 1193, Mabank, TX 75147
Puett's Antique Radio Topics, P.O. Box 28572, Dallas, TX 75228
Radio Age, 636 Cambridge Road, Augusta, GA 30909
Wireless Trader, 4290 Bells Ferry Road, Suite 106-36, Kennesaw, GA 30144

Clubs

There are two national clubs and dozens of regional clubs in the U.S. serving the antique radio collector. There are also clubs in Australia, Britain, Canada, France, Germany, Ireland, Italy, Japan, the Netherlands, New Zealand and Norway. The following is a complete listing of all American clubs at present.

Alabama Historical Radio Society, 4721 Overwood Circle, Birmingham, AL 35222

Antique Radio Club of America, c/o Jim and Barbara Rankin, 3445 Adaline Dr., Stow, OH 44224. National club with regional chapters in Hawaii, Kentucky and West Virginia.

Antique Radio Club of Illinois, c/o Carl Knipfel, Rt. 3, Veterans Rd., Morton, IL 61550. Hosts annual "Radiofest" each August in Elgin, Illinois.

Antique Radio Club of Schenectady, c/o Jack Nelson, 915 Sherman St., Schenectady, NY 12303.

Antique Radio Collectors of Ft. Smith, Arkansas, c/o Wanda Conatser, 7917 Hermitage Dr., Ft. Smith, AR 72903.

Antique Radio Collectors of Ohio, c/o Karl Koogle, 2929 Hazelwood Ave., Dayton, OH 45419.

Antique Wireless Association, P.O. Box "E", Breesport, NY 14816. National club. Hosts annual conference each September in Rochester, New York, which features the world's largest antique radio swap meet.

Arkansas Antique Radio Club, c/o Tom Burgess, P.O. Box 9769, Little Rock, AR 72219.

Arizona Antique Radio Club, 8311 E. Via de Sereno, Scottsdale, AZ 85258

Belleville Area Antique Radio Club, c/o Karl Stegman, 4 Cresthaven Dr., Belleville, IL 62221.

Buckeye Antique Radio and Phonograph Club, c/o Steve Dando, 4572 Mark Trail, Copley, OH 44321.

California Historical Radio Society, c/o Jim McDowell, 2265 Panoramic Dr., Concord, CA 94520.

California Historical Radio Society - North valley Chapter, c/o Norm Braithwaite, P.O. Box 2443, Redding, CA 96099.

Carolina Antique Radio Society, c/o Carl R. Shirley, 824 Fairwood Rd., Columbia, SC 29209.

Central Jersey Antique Radio Club, c/o Tony Flanagan, 92 Joysan Ter., Freehold, NJ 07728

Cincinnati Antique Radio Collectors, c/o Tom Ducro, 6805 Palmetto, Cincinnati, OH 45227.

Colorado Radio Collectors, c/o Bruce Young, 4030 Quitman St., Denver, CO 80212.

Connecticut Vintage Radio Collectors Club, c/o John C. Ellsworth, Vinatge Radio and Communications Museum of Connecticut, 665 Arch St., New Britain, CT 06051.

Delaware Valley Historical Radio Club, c/o Radio Attic, P.O. Box 624, Lansdale, PA 19446.

Florida Antique Wireless Group, c/o Paul Currie, Box 738, Chuluota, FL 32766.

Greater Boston Antique Radio Collectors, c/o Richard C. Foster, 12 Shawmut Ave., Cochituate, MA 01778.

Greater New York vintage Wireless Association, c/o Bob Scheps, 12 Garrity Ave., Ronkonkoma, NY 11779.

Hawaii Chapter, Antique Radio Club of America, 95-2044 Waikalani Place C-401, Mililani, HI 96789.

Hawaii Historical Radio Club, c/o Bob or Tina Wiepert, 98-1438C Koahehe St., Pearl City, HI 96782.

Houston Vintage Radio Association, P.O. Box 31276, Houston, TX 77231-1276.

Hudson Valley Antique Radio and Phonograph Society, c/o John Gramm, P.O. Box 1, Campbell Hall, NY 10916.

Kentucky Chapter, Antique Radio Club of America, 1907 Lynn Lea Rd., Louisville, KY 40216-2836.

Indiana Historical Radio Society, 245 N. Oakland Ave., Indianapolis, IN 46201-3360.

Michigan Antique Radio Club, c/o Larry Anderson, 3453 Balsam NE, Grand Rapids, MI 49505.

Mid-America Antique Radio Club, c/o Carleton Gamet, 2307 W. 131 St., Olathe, KS 66061.

Mid-Atlantic Antique Radio Club, c/o Joe Koester, 249 Spring Gap South, Laurel, MD 20724.

Middle Tennessee Old Radio Club, c/o Grant Manning, Rt. 2, Box 127A, Smithville, TN 37166.

Mid-South Antique Radio Collectors, Inc., c/o Ron Ramirez, 811 Maple St., Providence, KY 42450-1857.

Mississippi Historical Radio and Broadcasting Society, c/o Randy Guttery, 2412 C St., Meridian, MS 39301.

Music City Vintage Radio & Phonograph Society, P.O. Box 22291, Nashville, TN 37202

Nebraska Radio Collectors Antique Radio Club, c/o Steve Morton, 905 W. First, North Platte, NE 69101.

New England Antique Radio Club, c/o Marty Bunis, RR 1, Box 36, Bradford, NH 03221.

Niagara Frontier Wireless Association, c/o Gary Parzy, 135 Autumnwood, Cheektowaga, NY 14227.

Northland Antique Radio Club, P.O. Box 18362, Minneapolis, MN 55418.

Northwest Vintage Radio Society, P.O. Box 82379, Portland, OR 97282-0379.

Pittsburgh Antique Radio Society, Inc., c/o Richard J. Harris, Jr., 407 Woodside Rd., Pittsburgh, PA 15221.

Sacramento Historical Radio Society, P.O. Box 162612, Sacramento, CA 95816-9998.

Society for the Preservation of Antique Radio Knowledge, c/o WQRP Radio, 2673 S. Dixie Dr., Dayton, OH 45409.

Southeastern Antique Radio Society, c/o David Martin, 1502 Wood Thrush Way, Marietta, GA 30062.

Southern California Antique Radio Society, c/o C. Alan Smith, 6368 Charing St., San Diego, CA 92117.

Southern Vintage Wireless Association, c/o Bill Moore, 1005 Fieldstone Ct., Huntsville, AL 35803.

South Florida Antique Radio Collectors, c/o Victor Marett, 3201 N.W. 18 St., Miami, FL 33125.

Vintage Radio & Phonograph Society, c/o Larry Lamia, P.O. Box 165345, Irving, TX 75016.

Vintage Radio Unique Society, c/o Jerryl W. Sears, 312 Auburndale St., Winston-Salem, NC 27104.

Western Wisconsin Antique Radio Collectors Club, c/o Blake Nichols, Rt. 1, Box 182-A4, Stoddard, WI 54658.

West Virginia Chapter, Antique Radio Club of America, c/o Geoff Bourne, 405 8th Ave., St. Albans, WV 25177.

Parts and Supplies

This is a very small sampling of what is available to the antique radio collector.

Antique Audio (of Texas), 5555 N. Lamar, Suite H-105, Austin, TX 78751. General line of supplies, and speaker repair.

Antique Audio (of Michigan), 41560 Schoolcraft, Plymouth, MI 48170. Capacitors, other parts.

Antique Electronic Supply, 6221 S. Maple Ave., Tempe, AZ 85283. General line of supplies.

Antique Radio Service (Richard Foster), 12 Shawmut Ave., Cochituate, MA 01778. Specializing in cabinet repair and restoration.

Larry Bordonaro, 5744 Tobias, Van Nuys, CA 91411. Reproduction of plastic parts.

Hank Brazeal, 103 N. Lake Point Court, Crossville, TN 38555. Speaker repair.

Constantine's, 2050 Eastchester Rd., Bronx, NY 10461. Refinishing supplies and veneer.

Don Diers, 4276 North 50 St., Milwaukee, WI 53216. Tubes and other parts.

Electron Tube Enterprises, Box 8311, Essex, VT 05451. Tubes.

Frontier Electronics, Box 38, Lehr, ND 58460. Electrolytic capacitors sold and rebuilt.

Craig L. Graybar Furniture Works, Ltd., 1535 S. 84th St., West Allis, WI 53214. Cabinet repair and restoration.

Jackson Speaker Service, 217 Crestbrook Dr., Jackson, MI 49203. Speaker repair.

Michael Katz, 3987 Daleview Ave., Seaford, NY 11783. Grille cloth.

Lakes Loudspeaker Service, 4400 W. Hillsboro Blvd., Coconut Creek, FL 33073. Speaker repair.

Mohawk Finishing Products, Inc., Route 30 North, Amsterdam, NY 12010. Refinishing supplies.

John Okolowicz, 624 Cedar Hill Rd., Ambler, PA 19002. Grille cloth.

Olde Tyme Radio Company, 2445 Lyttonsville Rd., Suite 317, Silver Spring, MD 20910. General line of supplies.

Play Things Of Past (Gary Schneider), 3552 W. 105th St., Cleveland, OH 44111. General line of supplies.

Puett Electronics, P.O. Box 28572, Dallas, TX 75228. General line of supplies.

R.F. Perspectives (Victor Smith), 147-29 Hoover Ave., Briarwood, NY 11435. Tubes.

Sound Remedy, 331 Virginia Ave., Collingswood, NJ 08108. Speaker repair.

SRS Enterprises (The Speaker Shop), 318 S. Wahsatch Ave., Colorado Springs, CO 80903. Speaker repair.

Vintage TV & Radio, 3498 W. 105th St., Cleveland, OH 44111. General line of supplies.

Appendix II

Chapter 2

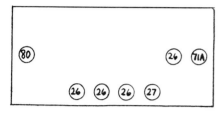

2-1. Models 511, 512, 513, 514, 515, 531, 551 & 571

Chapter 3

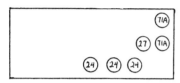

3-1. Model 40.

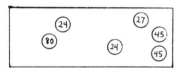

3-2. Model 65.

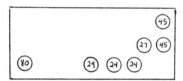

3-3. Model 76.

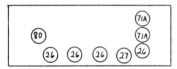

3-4. Model 86.

3-5. Model 87.

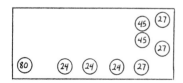

3-6. Model 95.

Chapter 4

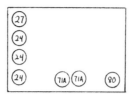

4-1. Model 20.

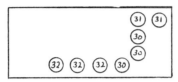

4-2. Model 30.

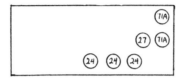

4-3. Model 41.

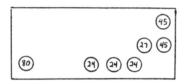

4-4. Model 77.

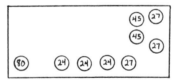

4-5. Models 96, 296, & Concert Grand.

Chapter 5

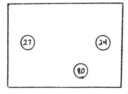

5-1. The Model 4 shortwave converter.

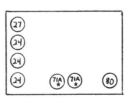

5-2. Model 21. * - 45 tubes in very late production.

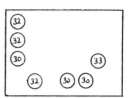

5-3. Model 35.

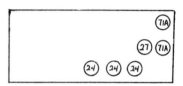

5-4. Model 42.

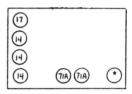

5-5. Model 46. * - ballast

5-6. Model 50.

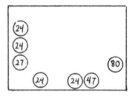

5-7. Model 70, 270, 370, & 570.

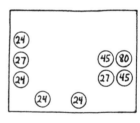

5-8. Model 90. Type 1 (June). Has Normal-Maximum switch in back.

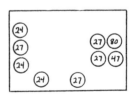

5-9. Model 90. Type 2 (October). Has Automatic Volume Control (AVC).

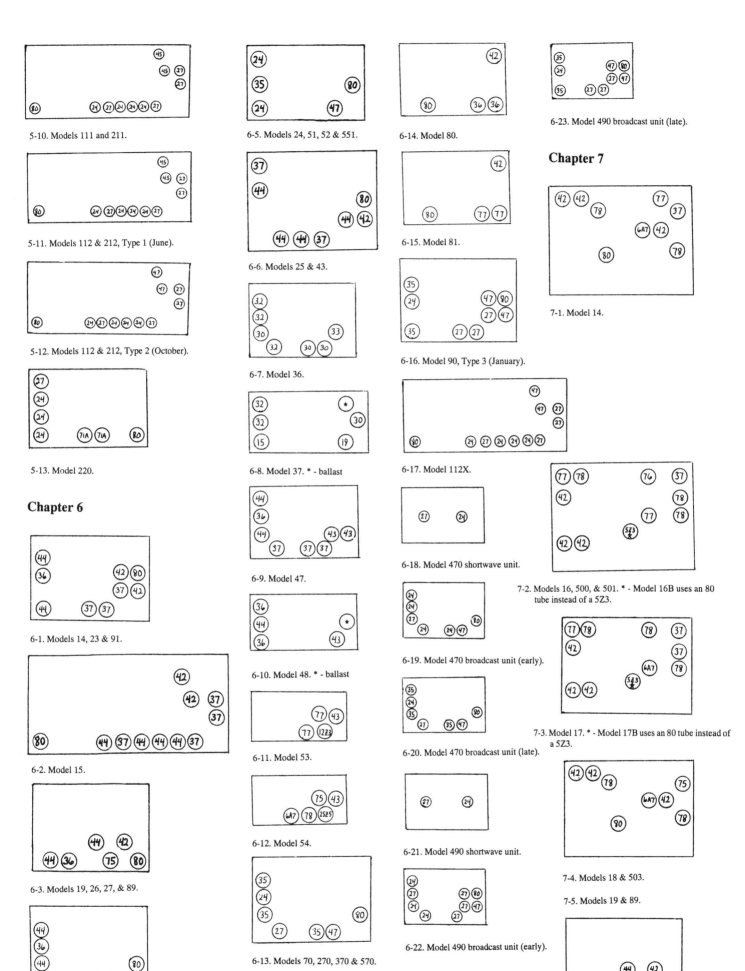

5-10. Models 111 and 211.

5-11. Models 112 & 212, Type 1 (June).

5-12. Models 112 & 212, Type 2 (October).

5-13. Model 220.

Chapter 6

6-1. Models 14, 23 & 91.

6-2. Model 15.

6-3. Models 19, 26, 27, & 89.

6-4. Models 22 & 71.

6-5. Models 24, 51, 52 & 551.

6-6. Models 25 & 43.

6-7. Model 36.

6-8. Model 37. * - ballast

6-9. Model 47.

6-10. Model 48. * - ballast

6-11. Model 53.

6-12. Model 54.

6-13. Models 70, 270, 370 & 570.

6-14. Model 80.

6-15. Model 81.

6-16. Model 90, Type 3 (January).

6-17. Model 112X.

6-18. Model 470 shortwave unit.

6-19. Model 470 broadcast unit (early).

6-20. Model 470 broadcast unit (late).

6-21. Model 490 shortwave unit.

6-22. Model 490 broadcast unit (early).

6-23. Model 490 broadcast unit (late).

Chapter 7

7-1. Model 14.

7-2. Models 16, 500, & 501. * - Model 16B uses an 80 tube instead of a 5Z3.

7-3. Model 17. * - Model 17B uses an 80 tube instead of a 5Z3.

7-4. Models 18 & 503.

7-5. Models 19 & 89.

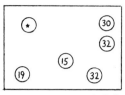

7-6. Models 38 & 38-A. * - ballast (38-A only)

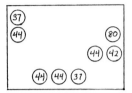

7-7. Model 43.

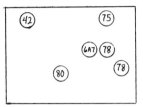

7-8. Models 44 & 504.

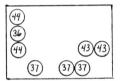

7-9. Model 47.

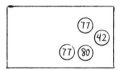

7-10. Models 57 & 58.

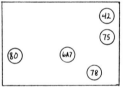

7-11. Models 60 & 505.

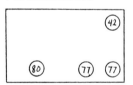

7-12. Model 84.

Chapter 8

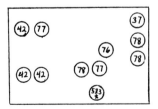

8-1. Models 16, 500 & 501. * - Model 16B uses an 80 tube instead of a 5Z3.

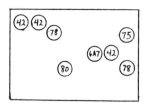

8-2. Models 18, 118, 503 & 507.

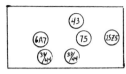

8-3. Model 28.

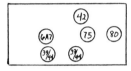

8-4. Models 29 & 45.

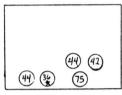

8-5. Model 32. * - 77 in later models. Separate power supply uses an 84 tube.

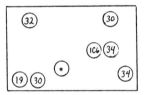

8-6. Models 34 & 34-A. * = ballast (34-A only)

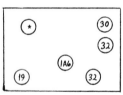

8-7. Models 38 & 38-A. * - ballast (38-A only)

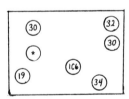

8-8. Models 39 & 39-A. * - ballast (39-A only)

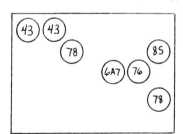

8-9. Model 49.

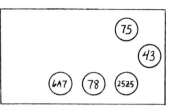

8-10. Model 54.

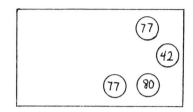

8-11. Model 59.

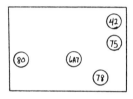

8-12. Models 60 & 505.

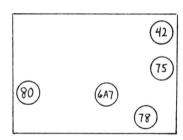

8-13. Model 66.

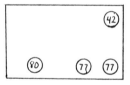

8-14. Model 84.

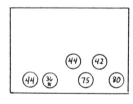

8-15. Model 89. * - 77 in later models.

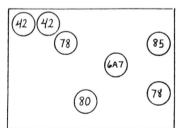

8-16. Model 97.

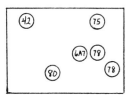

8-17. Models 144 & 506.

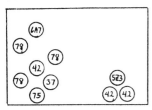

8-18. Model 200.

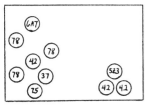

8-19. Models 201 & 509.

Chapter 9

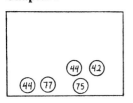

9-1. Model 32. Separate power supply uses an 84 tube.

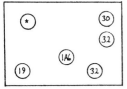

9-2. Models 38 & 38-A. * - ballast (38-A only)

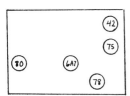

9-3. Model 60.

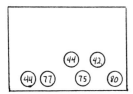

9-4. Model 89.

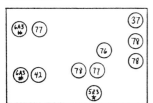

9-5. Model 116. * - 80 tube in 116B. # - 42 tubes in 116B.

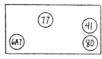

9-6. Model 600.

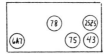

9-7. Model 602.

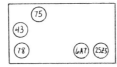

9-8. Model 604.

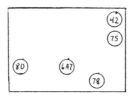

9-9. Model 610.

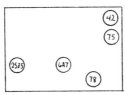

9-10. Model 611.

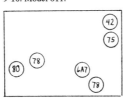

9-11. Models 620 & 625.

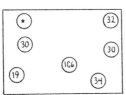

9-12. Models 623 & 623-A. * - ballast (623-A only)

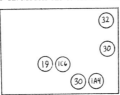

9-13. Model 624.

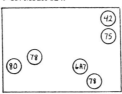

9-14. Models 630 & 635.

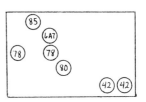

9-15. Models 640 & 645.

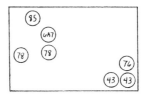

9-16. Model 641.

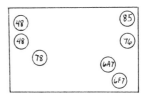

9-17. Model 642.

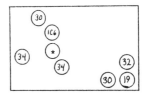

9-18. Models 643 & 643-A. * - ballast (643-A only)

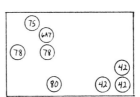

9-19. Models 650 & 655.

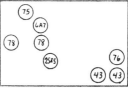

9-20. Model 651.

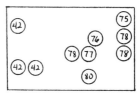

9-21. Models 660 & 665.

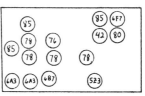

9-22. Model 680.

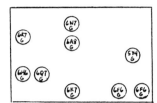

10-1. Model 37-9.

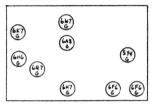

10-2. Model 37-10.

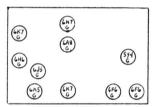

10-3. Model 37-11.

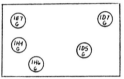

10-4. Model 37-33.

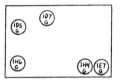

10-5. Model 37-34.

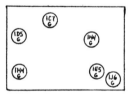

10-6. Model 37-38.

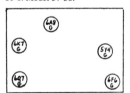

10-7. Model 37-60.

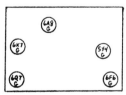

10-8. Model 37-61.

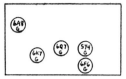

10-9. Model 37-62.

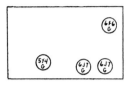

10-10. Model 37-84.

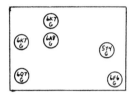

10-11. Model 37-89.

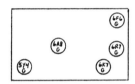

10-12. Model 37-93.

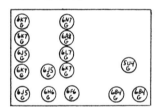

10-13. Model 37-116X.

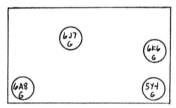

10-14. Model 37-600.

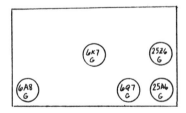

10-15. Model 37-602.

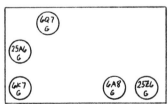

10-16. Model 37-604.

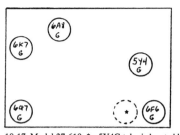

10-17. Model 37-610. * - 5Y4G tube is located here in Code 122 sets.

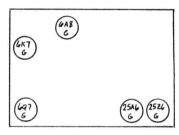

10-18. Model 37-611.

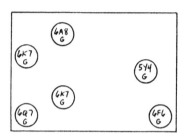

10-19. Model 37-620.

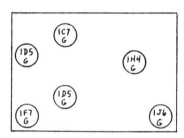

10-20. Model 37-623.

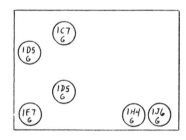

10-21. Model 37-624.

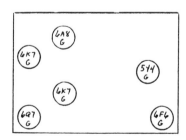

10-22. Model 37-630.

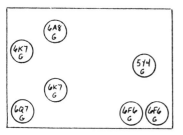

10-23. Model 37-640.

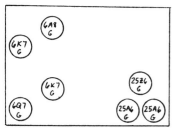

10-24. Model 37-641.

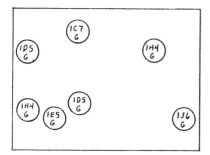

10-25. Model 37-643.

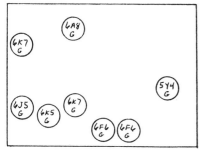

10-26. Model 37-650.

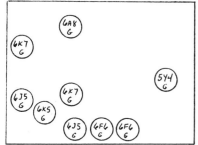

10-27. Model 37-660.

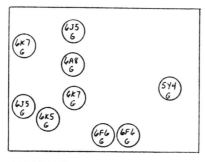

10-28. Model 37-665.

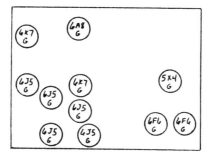

10-29. Model 37-670.

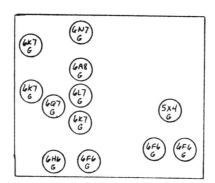

10-30. Model 37-675.

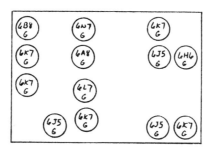

10-31. The Model 37-690 main chassis.

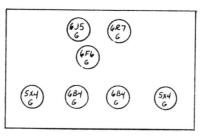

10-32. The Model 37-690 audio/power supply chassis.

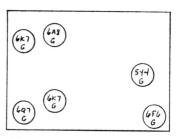

10-33. Model 37-2620.

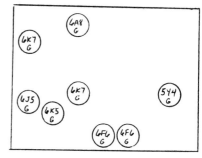

10-34. Model 37-2650.

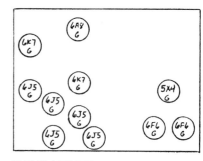

10-35. Model 37-2670.

Chapter 11

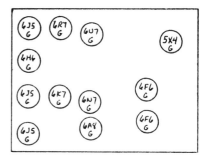

11-1. Model 38-1.

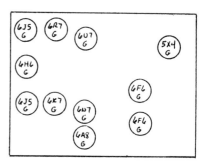

11-2. Model 38-2.

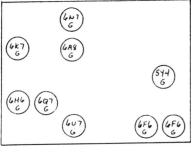

11-3. Model 38-3.

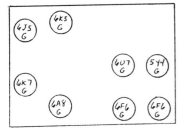

11-4. Model 38-4.

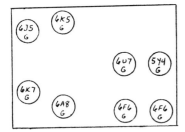

11-5. Model 38-5.

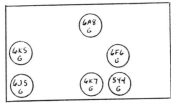

11-6. Model 38-7.

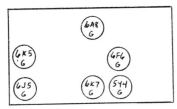

11-7. Models 38-8 & 38-9.

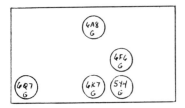

11-8. Model 38-10.

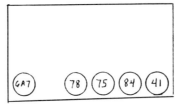

11-9. Model 38-12.

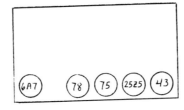

11-10. Model 38-14.

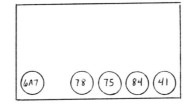

11-11. Model 38-15.

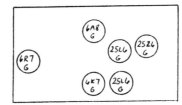

11-12. Model 38-22.

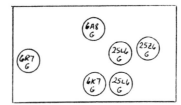

11-13. Model 38-23.

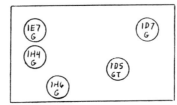

11-14. Model 38-33.

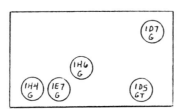

11-15. Model 38-34.

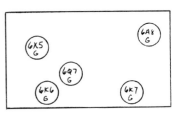

11-16. Model 38-35.

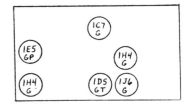

11-17. Model 38-38.

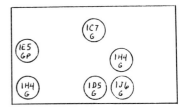

11-18. Model 38-39

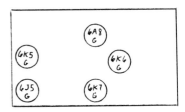

11-19. Model 38-40. Separate power supply uses a 6X5G tube.

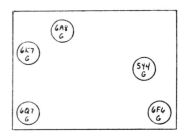

11-20. Model 38-60.

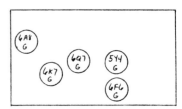

11-21. Model 38-62.

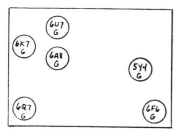

11-22. Model 38-89.

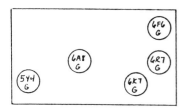

11-23. Model 38-93.

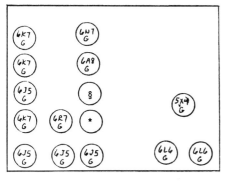

11-24. Model 38-116. * - 6K7G in Code 121, 6U7G in
Code 122. # -6L7G in Code 121, 6A8G in Code
122.

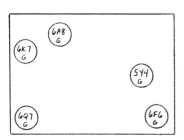

11-25. Model 38-610B.

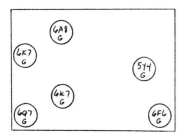

11-26. Model 38-620.

11-27. Model 38-623.

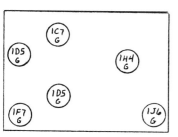

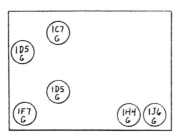

11-28. Model 38-624.

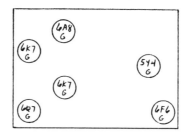

11-29. Model 38-630.

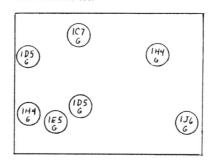

11-30. Model 38-643.

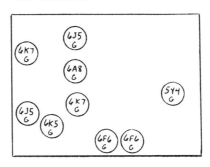

11-31. Model 38-665.

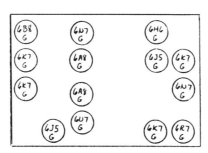

11-32. Model 38-690, the main chassis.

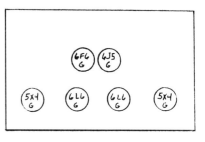

11-33. Model 38-690, the audio/power supply chassis.

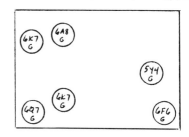

11-34. Models 38-2620 & 38-2630.

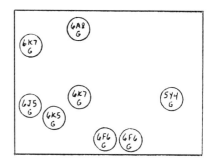

11-35. Model 38-2650.

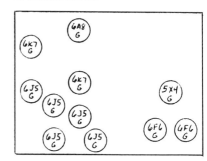

11-36. Model 38-2670.

Chapter 12

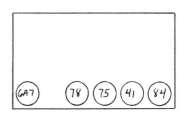

12-1. Models 39-6 & 39-7.

175

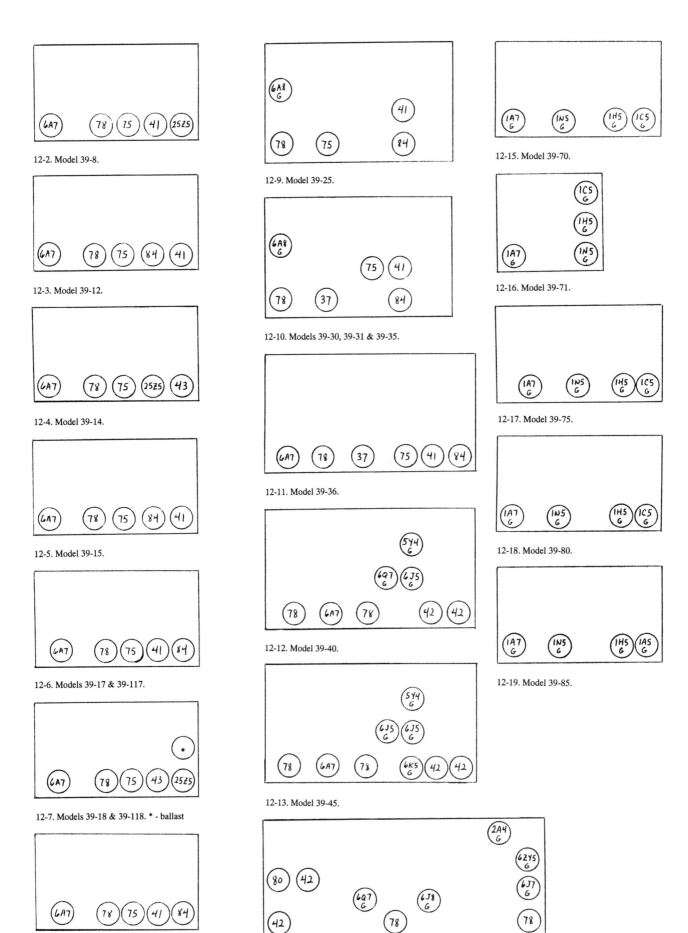

12-2. Model 39-8.

12-3. Model 39-12.

12-4. Model 39-14.

12-5. Model 39-15.

12-6. Models 39-17 & 39-117.

12-7. Models 39-18 & 39-118. * - ballast

12-8. Models 39-19 & 39-119.

12-9. Model 39-25.

12-10. Models 39-30, 39-31 & 39-35.

12-11. Model 39-36.

12-12. Model 39-40.

12-13. Model 39-45.

12-14. Model 39-55. Separate remote unit uses a 30 tube.

12-15. Model 39-70.

12-16. Model 39-71.

12-17. Model 39-75.

12-18. Model 39-80.

12-19. Model 39-85.

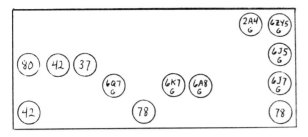

12-20. Model 39-116RX. Separate remote unit uses a 30 tube.

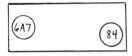

12-21. Models RP-l, RP-2, RP-3 & RP-4.

Chapter 13

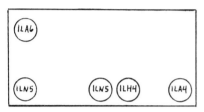

13-1. Model 40-84.

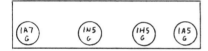

13-2. Model 40-110.

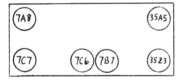

13-3. Models 40-115 & 40-124.

13-4. Models 40-120 & 40-125.

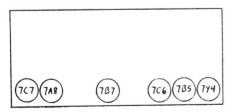

13-5. Models 40-130, 40-135, 40-170, 40-503, 40-506, 40-525, 40-526 & 40-527.

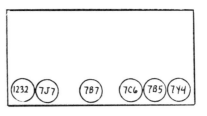

13-6. Models 40-140, 40-145 & 40-507.

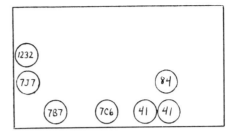

13-7. Models 40-150 & 40-180.

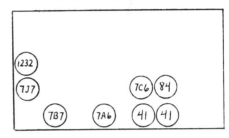

13-8. Models 40-155, 40-185 & 40-190.

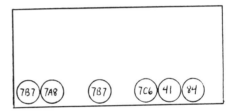

13-9. Model 40-158.

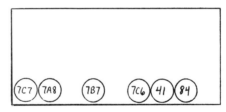

13-10. Model 40-160.

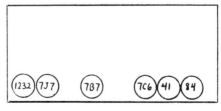

13-11. Model 40-165.

13-12. Model 40-195 & Model 40-201, Code 122.

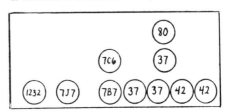

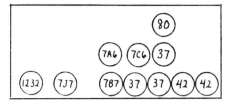

13-13. Model 40-200 & Model 40-201, Code 121.

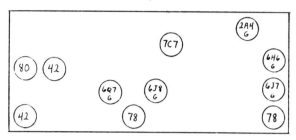

13-14. Models 40-205 & 40-510. Separate remote unit uses a 30 tube.

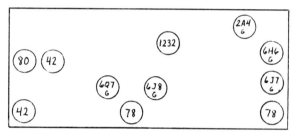

13-15. Model 40-510P. Separate remote unit uses a 30 tube.

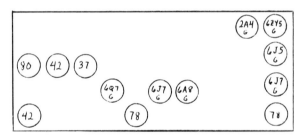

13-16. Models 40-216 & 40-516. Separate remote unit uses a 30 tube.

Chapter 14

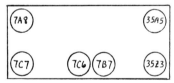

14-1. Models 41-22CL, 41-220 & 41-225.

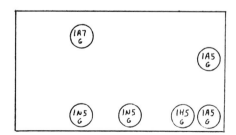

14-2. Model 41-110.

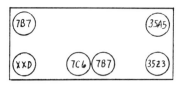

14-3. Models 41-221, 41-226 & 41-231.

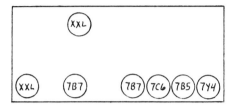

14-4. Models 41-230 & 41-235.

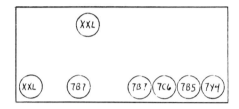

14-5. Model 41-240.

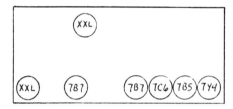

14-6. Models 41-245 & 41-246.

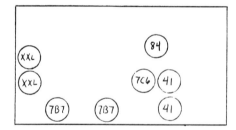

14-7. Model 41-250.

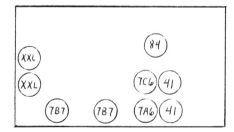

14-8. Models 41-255 & 41-256.

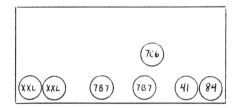

14-9. Model 41-260.

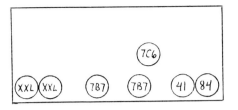

14-10. Model 41-265.

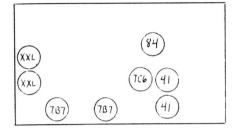

14-11. Model 41-280.

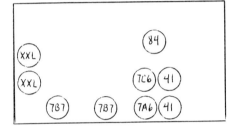

14-12. Models 41-285 & 41-287.

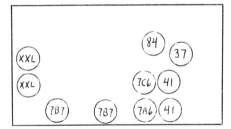

14-13. Model 41-290.

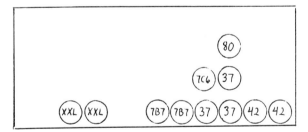

14-14. Model 41-295.

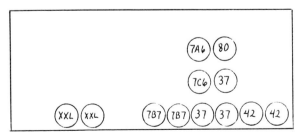

14-15. Models 41-300 & 41-315.

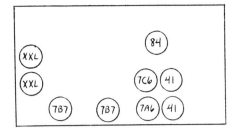

14-16. Model 41-296.

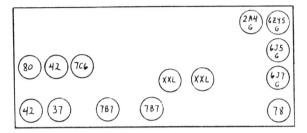

14-17. Model 41-316. Separate remote unit uses a 30 tube.

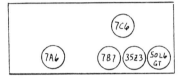

14-18. Model 41-601.

14-19. Model 41-602.

14-20. Model 41-603.

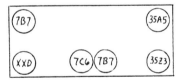

14-21. Models 41-604, 41-605, & 41-607.

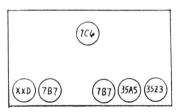

14-23. Models 41-610 & 41-611.

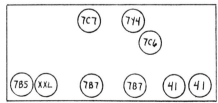

14-22. Models 41-608 & 41-609.

179

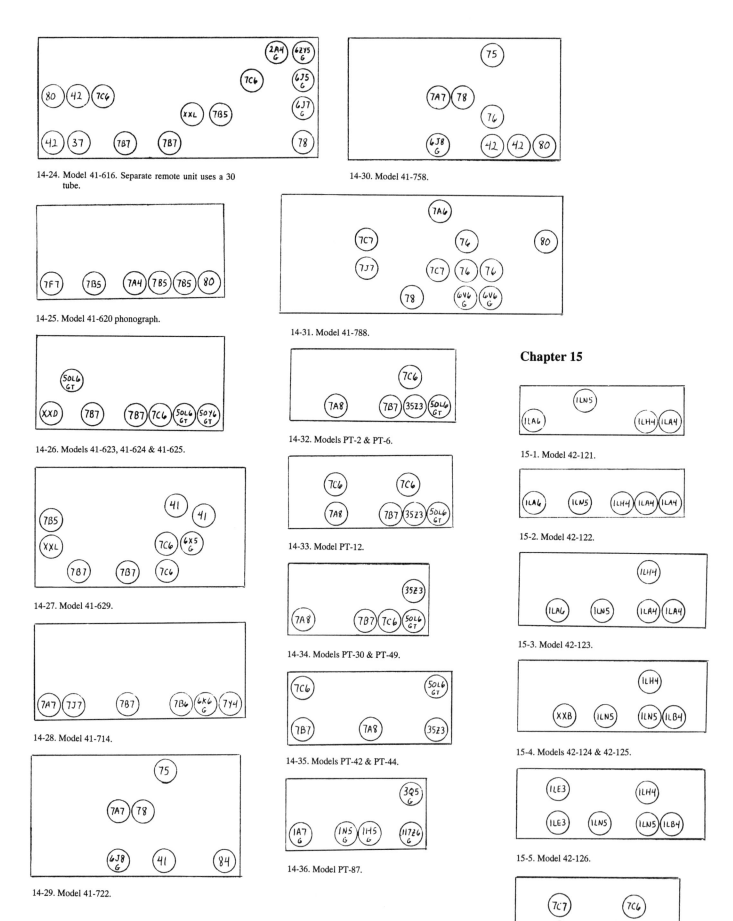

14-24. Model 41-616. Separate remote unit uses a 30 tube.

14-30. Model 41-758.

14-25. Model 41-620 phonograph.

14-31. Model 41-788.

14-26. Models 41-623, 41-624 & 41-625.

14-32. Models PT-2 & PT-6.

Chapter 15

14-27. Model 41-629.

14-33. Model PT-12.

15-1. Model 42-121.

15-2. Model 42-122.

14-28. Model 41-714.

14-34. Models PT-30 & PT-49.

15-3. Model 42-123.

14-35. Models PT-42 & PT-44.

15-4. Models 42-124 & 42-125.

14-29. Model 41-722.

14-36. Model PT-87.

15-5. Model 42-126.

15-6. Models 42-321 & 42-PT10.

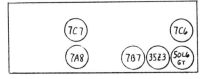

15-7. Model 42-322.

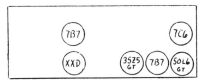

15-8. Model 42-323. Code 122 sets use a 35Z3 tube in place of the 35Z5GT tube.

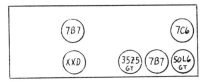

15-9. Model 42-327. Code 122 sets use a 35Z3 tube in place of the 35Z5GT tube.

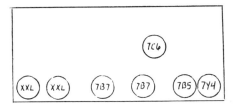

15-10. Model 42-335.

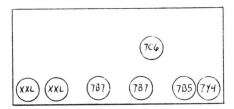

15-11. Model 42-340.

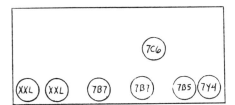

15-12. Model 42-345.

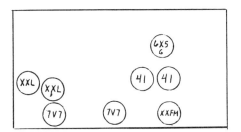

15-13. Models 42-355 & 42-390.

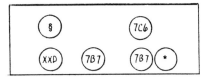

15-14. Model 42-358. * - 50L6GT in Code 121, 35L6GT in Code 122. # - 35Z3 in Code 121, 50Y6GT in Code 122.

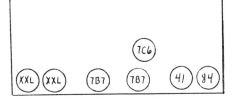

15-15. Model 42-360. Some models use a 7Y4 tube in place of the 84 tube.

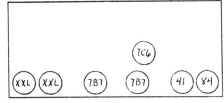

15-16. Model 42-365.

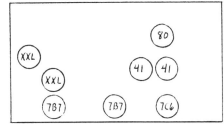

15-17. Model 42-380.

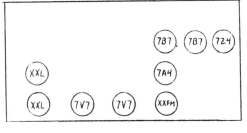

15-18. Model 42-395.

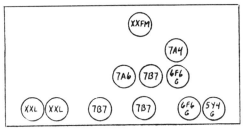

15-19. Model 42-400.

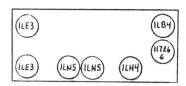

15-20. Models 42-853 & 42-854.

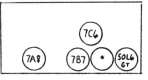

15-21. Model 42-1001. * - 35Z3 in Code 121, 50Y6GT in Code 122

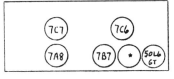

15-22. Model 42-1002. * - 35Z3 in Code 121, 50Y6GT in Code 122

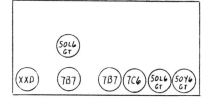

15-23. Model 42-1003.

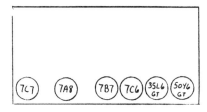

15-24. Model 42-1004.

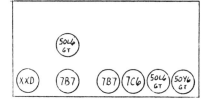

15-25. Model 42-1005.

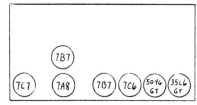

15-26. Model 42-1006.

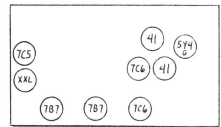

15-27. Models 42-1008 & 42-1009.

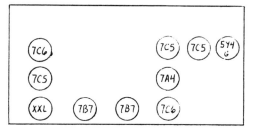

15-28. Models 42-1010 & 42-1011.

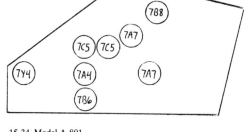

15-34. Model A-801.

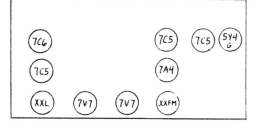

15-29. Models 42-1012 & 42-1013.

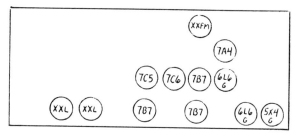

15-30. Model 42-1015.

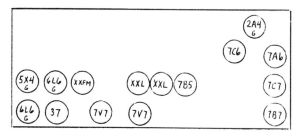

15-31. Model 42-1016. Separate remote unit uses a 30 tube.

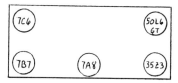

15-32. Models 42-KR3 & 42-KR5.

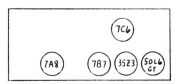

15-33. Models 42-PT2, 42-PT4 & 42-PT7.

Bibliography

Ghirardi, Alfred A. *Modern Radio Servicing*. New York: Radio & Technical Publishing Co., 1935.

Graf, Rudolf F. *Electronics Learning Dictionary*. Indianapolis: Howard W. Sams & Co., 1962.

Grinder, Robert E., and George H. Fathauer. *The Radio Collector's Directory and Price Guide*. Scottsdale, AZ: Ironwood Press, 1986.

Heimstead, Doug. "The Philco Predictas." *Antique Radio Classified*, Vol. 5, No. 12, December 1988, 4-8.

Houston, C. Douglas, Jr. 1942 Philco automatic record changers. Ortonville, MI, 1992.

Langley, Ralph H. *Radio Collector's Guide: 1921-1932*. Edited by Morgan E. McMahon. Palos Verdes Peninsula, CA: Vintage Radio, 1973.

Marcus, Abraham, William Marcus, and Ralph E. Horton. *Radio for Beginners*. New York: Prentice-Hall, Inc., 1943.

McMahon, Morgan E. *A Flick Of The Switch: 1930-1950*. North Highlands, CA: Vintage Radio, 1975.

Philadelphia Storage Battery Company. *Philco Radio Manual of Useful Information 1928-1929*. Philadelphia, 1928.

Philadelphia Storage Battery Company, brochure (1929 models). Philadelphia, 1929.

Philadelphia Storage Battery Company, dealer catalog (1930 models). Philadelphia, 1930.

Philadelphia Storage Battery Company. *Philco News*, Vol. I, No. 12. Philadelphia, October 1930.

Philadelphia Storage Battery Company. *Philco News*, vol. II, No. 1. Philadelphia, November 1930.

Philadelphia Storage Battery Company, brochure (1931 models). Philadelphia, 1931.

"Philco." *The Horn Speaker*, October 1979, 1-2. (First published by Philco-Ford Corporation, News Department, 1969.)

Philco Corporation, brochure (1941 models). Philadelphia, 1940.

Philco Corporation, dealer catalog (1942 models). Philadelphia, 1941.

Philco Radio & Television Corporation, brochure (1932-33 models). Philadelphia, 1932.

Philco Radio & Television Corporation. *Philco Book of Facts*. Philadelphia, 1932.

Philco Radio & Television Corporation, brochure (1933-34 models). Philadelphia, 1933.

Philco Radio & Television Corporation. *Philco Essential Service Data on All Models*. Philadelphia, 1934.

Philco Radio & Television Corporation. *Wiring Diagrams and Parts Lists, Philco Balanced-Unit Radio Receivers*. Philadelphia, 1934.

Philco Radio & Television Corporation, salesman's brochure (1935 models). Philadelphia, 1934.

Philco Radio & Television Corporation, brochure (1935 models). Philadelphia, 1935.

Philco Radio & Television Corporation. *Philco Service Bulletin No. 216 - Cabinet Repair Parts Lists for 1934-5 Models*. Phildelphia, 1935.

Philco Radio & Television Corporation, salesman's brochure (1936 models). Philadelphia, 1935.

Philco Radio & Television Corporation, dealer catalog (1937 models). Philadelphia, 1936.

Philco Radio & Television Corporation, brochure (1937 models). Philadelphia, 1937.

Philco Radio & Television Corporation. *Philco 1937 List Price Catalog*. Philadelphia, 1937.

Philco Radio & Television Corporation, dealer catalog (1938 models). Philadelphia, 1937.

Philco Radio & Television Corporation, dealer catalog (1939 models). Philadelphia, 1938.

Philco Radio & Television Corporation, brochure (1940 models). Philadelphia, 1939.

Philco Radio & Television Corporation, brochure (1940 Philco Transitone models). Philadelphia, 1940.

Philips Consumer Electronics Company. *The History of Philco: A Tradition of American Know-How and Value* (brochure). Knoxville, TN, 1982.

Poster, Harry. "Early Transistor TVs." *Antique Radio Classified*, Vol. 5, No. 7, July 1988, 6-7.

Rider, John F. *Perpetual Trouble Shooter's Manual*, vol. I-XIV. New York: John F. Rider Publisher, 1932-42.

Science and Invention, December 1930, 727 & 759.

U.S. Patent Office. Design patents for various Philco cabinets. Des. 81,426, 82,704, 83,956, 84,733, 85,473, 85,520, 85,818, 85,819, 86,440, 87,769, 101,061, 101,062, 101,063, 101,064, 101,087, 126,606, 126,863, 126,864, 127,155, 183,779, 183,780, 183,782, 186,335. Photocopy.

Wolkonowicz, John P. "The Philco Corporation: Historical Review & Strategic Analysis, 1892-1961." Master's thesis, Massachusetts Institute of Technology, 1981.

Index

RARITY SCALE

The following information describing the relative degree of rarity among Philco radio receivers made between 1928 and 1942 is furnished at the request of the publisher.

This guide is based on the author's personal experience. It should be remembered that conditions may vary in different geographic areas, and that relative degree of rarity is not necessarily an indicator of price.

Comments are welcome; however, the author assumes no responsibility for any inaccuracies in this rarity scale.

Degree of rarity is given as follows:

A - fairly hard to find
B - somewhat difficult to find
C - a set that is fairly easy to find

An asterisk (*) next to an A in the relative rarity column indicates the set may be rare.

Model	Cabinet	Relative Rarity
1928		
511	Metal table model	C
512	Metal table model	B
513	Metal table model	B
514	Metal table model	B
515	Metal table model	A*
531	Console	B
551	Highboy	A
571	Highboy radio-phono	A*
1929		
40	Console	A
40	Lowboy	A
40	Highboy	A
40	Deluxe highboy	A
65	Metal table model	C
65	Lowboy	C
65	Highboy	B
65	Deluxe highboy	A
76	Metal table model	C
76	Console	C
76	Lowboy	C
76	Highboy	B
76	Deluxe highboy	A
86	Console	C
86	Highboy	A
87	Lowboy	C
87	Highboy	B
87	Deluxe highboy	A
95	Metal table model	C
95	Lowboy	B
95	Highboy	A
95	Deluxe highboy	A
1930		
20	Cathedral (plain)	C
20	Cathedral (deluxe)	C
20	Consolette	C
30	Lowboy	A
30	Highboy	A
41	Console	A
41	Lowboy	A
41	Highboy	A
77	Metal table model	C
77	Console	C
77	Lowboy	C
96	Metal table model	C
96	Lowboy	C
96	Highboy	B
296	Lowboy radio-phono	B
296	Concert Grand	A*
1931		
4	Table model SW converter	B
21	Cathedral	B

Model	Cabinet	Relative Rarity
35	Cathedral	A
35	Highboy	A
42	Console	A
42	Lowboy	A
42	Highboy	A
46	Cathedral	A
46	Highboy	A
50	Cathedral	C
50	Lowboy	A
70	Cathedral	C
70	Highboy	B
90	Cathedral	C
90	Lowboy	B
90	Highboy	B
111	Lowboy	A
111	Highboy	A
112	Lowboy	B
112	Highboy	A
211	Lowboy radio-phono	A*
212	Lowboy radio-phono	A*
220	Console radio-phono	A*
270	Lowboy radio-phono	A
370	Chairside	B
570	Grandfather Clock	B
1932-33		
14LZX	Chairside w/separate speaker	B
15X	Console	B
15DX	Console	A
19B	Cathedral	B
19L	Lowboy	B
19LZ	Chairside	A*
19LZX	Chairside w/separate speaker	A
22L	Lowboy radio-phono	A*
23X	Console radio-phono	A*
24L	Lowboy radio-phono	A*
25L	Lowboy radio-phono	A*
26L	Lowboy radio-phono	A*
27L	Lowboy radio-phono	A*
36B	Cathedral	A
36L	Lowboy	A
36D	Highboy w/doors	A
37C	Table model	A
37L	Lowboy	A
43B	Cathedral	A
43H	Highboy	A
43X	Console	A
47B	Cathedral	A
47H	Highboy	A
47D	Highboy w/doors	A
47X	Console	A
48C	Table model	A
48L	Lowboy	A
51B	Cathedral	B
51L	Lowboy	A
52B	Cathedral	A
52C	Table model	A
52L	Lowboy	A
53C	Table model	B
54C	Table model	C
70	Cathedral	C
70	Highboy	B
71B	Cathedral	B
71L	Lowboy	A
71H	Highboy	A
71D	Highboy w/doors	A
71LZ	Chairside	A
80	Colonial Clock	A
80B	Cathedral	C
80C	Table model	B
81B	Cathedral	B
89B	Cathedral	C
89L	Lowboy	A
90B	Cathedral	C
90X	Console	B
91B	Cathedral	B
91L	Lowboy	A
91D	Highboy w/doors	A
91X	Console	B
112X	Console	B
270	Lowboy radio-phono	A
370	Chairside	B

470	Lowboy AM/SW	A
490	Lowboy AM/SW	A
551	Colonial Clock	A
570	Grandfather Clock	A

1933-34

14B	Cathedral	A
14L	Lowboy	A
14X	Console	A
14MX	Console	A
14RX	Chairside w/separate speaker	A
16B	Cathedral	B
16B	Tombstone	B
16L	Lowboy	B
16X	Console	B
16RX	Chairside w/separate speaker	B
17B	Cathedral	A
17L	Lowboy (early version)	A
17L	Lowboy (late version)	A
17D	Highboy w/doors	A
17X	Console	A
17RX	Chairside w/separate speaker	A
18B	Cathedral	A
18L	Lowboy	A
18H	Highboy (early version)	A
18H	Highboy (late version)	B
18D	Highboy w/doors	A
18RX	Chairside w/separate speaker	A
18X	Console	A
19B	Cathedral	B
19H	Highboy	B
19TX	Chairside w/separate speaker	A*
38B	Cathedral (early version)	C
38B	Cathedral (late version)	C
38L	Lowboy	B
43B	Cathedral	A
43H	Highboy	A
44B	Cathedral	A
44H	Highboy	A
47B	Cathedral	A
47D	Highboy w/doors	A
47H	Highboy	A
47X	Console	A
54C	Table model	C
57C	Table model	C
58C	Table model	B
60B	Cathedral (early version)	C
60B	Cathedral (late version)	C
60MB	Tombstone	C
60L	Lowboy	A
84B	Cathedral	C
89B	Cathedral	C
89L	Lowboy	A
500X	Console radio-phono	A*
501X	Console radio-phono	A*
503L	Lowboy radio-phono	A*
504L	Lowboy radio-phono	A*
505L	Lowboy radio-phono	A*

1935

16B	Tombstone (early version)	B
16B	Tombstone (late version)	B
16L	Lowboy	B
16X	Console	A
16RX	Chairside w/separate speaker	A
18B	Cathedral	A
18B	Tombstone	A
18H	Highboy	A
18MX	Console	A
28C	Table model	B
28L	Lowboy	A
28D	Highboy w/doors	A
28F	Console	A
28CSX	Chairside	A*
29X	Console (early version)	A
29X	Console (late version)	B
29CSX	Chairside	A*
29TX	Chairside w/separate speaker (early version)	A
29TX	Chairside w/separate speaker (late version)	A
45C	Table model	B
45L	Lowboy	B
45F	Console	A

32B	Cathedral	A
32B	Tombstone	A
32L	Lowboy	A
34B	Cathedral	A
34B	Tombstone	A
34L	Lowboy	A
38B	Cathedral	C
38L	Lowboy	A
39B	Cathedral	B
39F	Console	A
49B	Cathedral	A
49B	Tombstone	A
49D	Highboy w/doors (early version)	A
49D	Highboy w/doors (late version)	A
49H	Highboy	A
49X	Console (early version)	A
49X	Console (late version)	A
54C	Table model	C
54S	Table model	B
59C	Table model	B
59S	Table model	B
60B	Cathedral	C
60L	Lowboy	A
66B	Cathedral	B
66B	Tombstone	B
66S	Tombstone	A
66L	Lowboy	A
84B	Cathedral (early version)	C
84B	Cathedral (late version)	B
89B	Cathedral	B
89L	Lowboy	A
97B	Tombstone	A*
97X	Console	A*
118B	Cathedral	B
118B	Tombstone	B
118D	Highboy w/doors (early version)	A
118D	Highboy w/doors (late version)	A
118H	Highboy	A
118X	Console (early version)	A
118X	Console (late version)	A
118MX	Console	A
118RX	Chairside w/separate speaker	A
144B	Cathedral	B
144B	Tombstone	B
144H	Highboy	A
144X	Console (early version)	A
144X	Console (late version)	A
200X	Console	A*
201X	Console	A*
500X	Console radio-phono	A*
501X	Console radio-phono	A*
503L	Lowboy radio-phono	A*
505L	Lowboy radio-phono	A*
506L	Lowboy radio-phono	A*
507L	Lowboy radio-phono	A*
509X	Console radio-phono	A*

1936

32B	Tombstone	A
32F	Console	A
38B	Cathedral (early version)	C
38B	Cathedral (late version)	C
38F	Console	A
60B	Cathedral (early version)	C
60B	Cathedral (late version)	C
60F	Console	A
89B	Cathedral	B
89F	Console	A
116B	Tombstone (early version)	B
116B	Tombstone (late version)	A
116X	Console	B
116PX	Console radio-phono	A
600C	Table model	A
602C	Table model	A
604C	Table model	A
610B	Tombstone	C
610T	Table model	B
610F	Console	B
611B	Tombstone	A
611F	Console	A
620B	Tombstone	C
620F	Console	B
623B	Tombstone (early version)	B
623B	Tombstone (late version)	A
623F	Console	A

624B	Tombstone	A
624F	Console	A
625B	Tombstone	B
625J	Console	B
630B	Tombstone	B
630X	Console	B
630PF	Console radio-phono	A
630CSX	Chairside	A
635B	Tombstone	B
635J	Console	A
635CSX	Chairside	A
640B	Tombstone	B
640X	Console	A
641B	Tombstone	A
641X	Console	A
642B	Tombstone	A
642F	Console	A
643B	Tombstone	A
643X	Console	A
645B	Tombstone	A
645X	Console	A
650B	Tombstone	B
650X	Console	B
650MX	Console	A
650H	Console	A
650PX	Console radio-phono	A
650RX	Chairside w/separate speaker	A*
651B	Tombstone	A
651	Console	A
655B	Tombstone	B
655H	Console	A
655X	Console	B
655MX	Console	A
655PX	Console radio-phono	A
655RX	Chairside w/separate speaker	A*
660L	Console	A
660X	Console	A
665L	Console	A
665X	Console	A
680X	Console (early version)	A*
680X	Console (late version)	A*

1937

37-9X	Console	B
37-10X	Console	B
37-11X	Console	B
37-33B	Cathedral (early version)	B
37-33B	Cathedral (late version)	A
37-33F	Console	B
37-34B	Cathedral	A
37-34B	Tombstone	A
37-34F	Console	A
37-38B	Tombstone	B
37-38F	Console	B
37-38J	Console	A
37-60B	Cathedral	C
37-60B	Tombstone	A
37-60F	Console	B
37-61B	Cathedral	C
37-61F	Console	B
37-62C	Table model	B
37-84B	Cathedral (early version)	C
37-84B	Cathedral (late version)	A
37-89B	Cathedral	B
37-89F	Console	B
37-93B	Cathedral	C
37-116X	Console w/shadow meter, normal tuning	A*
37-116X	Console "De Luxe" w/automatic tuning	A
37-600C	Table model	B
37-602C	Table model	B
37-604C	Table model	B
37-610B	Tombstone	C
37-610T	Table model	C
37-610J	Console	B
37-611B	Tombstone	A
37-611T	Table model	A
37-611F	Console	A
37-611J	Console	A
37-620B	Tombstone	C
37-620T	Table model	A
37-620J	Console	B
37-623B	Tombstone	B
37-623J	Console	A

37-624B	Tombstone	A
37-624J	Console	A
37-630T	Table model	C
37-630X	Console	B
37-640B	Tombstone	B
37-640X	Console	B
37-640MX	Console	B
37-641B	Tombstone	A
37-641X	Console	A
37-643B	Tombstone	A
37-643X	Console	A
37-650B	Tombstone	B
37-650X	Console	B
37-660B	Tombstone	B
37-660X	Console, walnut	B
37-660X	Console, mahogany	A
37-665B	Tombstone	A
37-665X	Console	A
37-670B	Tombstone	B
37-670X	Console	B
37-675X	Console w/shadow meter, normal tuning	A*
37-675X	Console "De Luxe" w/automatic tuning	B
37-690X	Console	A*
37-2620B	Tombstone	B
37-2620J	Console	A
37-2650B	Tombstone	B
37-2650X	Console	A
37-2670B	Tombstone	B
37-2670X	Console	A

1938

38-1XX	Console	B
38-2XX	Console	B
38-3-3PC	Console radio-phono	A
38-3XX	Console	B
38-4XX	Console	C
38-5B	Tombstone	A
38-5X	Console	B
38-7T	Table model	B
38-7CS	Chairside	A
38-7XX	Console	C
38-8X	Console	B
38-9T	Table model	C
38-9K	Console	B
38-9	Console radio-phono	A
38-10T	Table model	C
38-10F	Console	A
38-12C	Table model	C
38-12CI	Table model	A
38-12T	Table model	C
38-12CB	Bakelite table model, brown	B
38-12CBI	Bakelite table model, ivory	A
38-14CB	Bakelite table model, brown	B
38-14CBI	Bakelite table model, ivory	A
38-14CS	Chairside	A
38-14T	Table model	C
38-15CB	Bakelite table model, brown	B
38-15CBI	Bakelite table model, ivory	A
38-15CS	Chairside	A
38-15T	Table model	C
38-22T	Table model	B
38-22CS	Chairside	A
38-22XX	Console	B
38-23K	Console	B
38-23T	Table model	B
38-23X	Console	B
38-33B	Tombstone	A
38-34B	Tombstone	A
38-34	Console	A
38-35B	Tombstone	A
38-35	Console	A
38-38T	Table model	C
38-38K	Console	A
38-38X	Console	A
38-39T	Table model	A
38-39K	Console	A
38-39X	Console	A
38-40T	Table model	A
38-40K	Console	A
38-40X	Console	A
38-60B	Tombstone	A
38-60F	Console	A
38-62C	Table model	B

38-62F	Console	A
38-89B	Tombstone	A
38-89K	Console	A
38-93B	Tombstone (early version)	A
38-93B	Tombstone (late version)	B
38-116XX	Console	A
38-610B	Tombstone	A
38-610J	Console	A
38-620T	Table model	B
38-623T	Table model	A
38-623K	Console	A
38-630K	Console	B
38-643B	Tombstone	A
38-643X	Console	A
38-665B	Tombstone	A
38-665X	Console	B
38-690XX	Console	A*
38-2620T	Table model	A
38-2630K	Console	A
38-2650B	Tombstone	A
38-2650X	Console	A
38-2670B	Tombstone	B
38-2670X	Console	A

1939

39-6C	Table model	B
39-6CI	Table model, ivory	B
39-7C	Table model	B
39-7T	Table model	A
39-8T	Table model	B
39-10RP	Table model phono	A
39-12CB	Bakelite table model, brown	A
39-12CBI	Bakelite table model, ivory	A
39-12T	Table model	A
39-12TP	Table model radio-phono	A
39-14CB	Bakelite table model, brown	A
39-14CBI	Bakelite table model, ivory	A
39-14T	Table model	A
39-15CB	Bakelite table model, brown	A
39-15CBI	Bakelite table model, ivory	A
39-15T	Table model	A
39-17F	Console	A
39-17T	Table model	B
39-18F	Console	A
39-18T	Table model	B
39-19F	Console	A
39-19PA	Console radio-phono	A
39-19PF	Console radio-phono	A
39-19PCS	Chairside radio-phono	A
39-19T	Table model	B
39-19TP	Table model radio-phono	A
39-25T	Table model	C
39-25XF	Console	B
39-30T	Table model	C
39-30PCX	Console radio-phono	A
39-31XF	Console	B
39-31XK	Console	A
39-3-31PA	Console radio-phono	A
39-35XX	Console	B
39-3-35PC	Console radio-phono	A
39-36XX	Console	B
39-40XX	Console	B
39-40PCX	Console radio-phono	A
39-2-40PC	Console radio-phono	A
39-45XX	Console	B
39-50RX	Console	A*
39-55RX	Console	A
39-70B	Tombstone	C
39-70F	Console	A
39-71T	Portable	A
39-75F	Console	A
39-75T	Table model	A
39-80B	Tombstone	B
39-80XF	Console	A
39-85B	Tombstone	B
39-85XF	Console	A
39-116PCX	Console radio-phono	A
39-116RX	Console	A
39-117F	Console	A
39-117T	Table model	A
39-118F	Console	A
39-118T	Table model	A
39-119F	Console	A
39-119T	Table model	A
907F	Consolette phono	A

907T	Table model phono	A
RP-1	Table model phono	A
RP-2	Table model phono	A
RP-3	Table model phono	A
RP-4	Consolette phono	A
TH-1	Table model	A
TH-3	Table model	A
TH-4	Table model	A
TH-4I	Table model, ivory	A
TP-4	Table model	A
TP-4I	Table model, ivory	A
TH-5	Table model	A
TH-5I	Table model, ivory	A
TP-5	Table model	A
TP-5I	Table model	A
TP-10	Table model	A
TP-11	Table model	A
TP-12	Table model	A

1940

40-74	Portable	B
40-81T	Portable	B
40-82T	Portable	A
40-84T	Portable	A
40-88T	Portable	B
40-90CB	Table model	B
40-95T	Table model	B
40-95F	Console	A
40-100	Table model	A
40-100	Console	A
40-105	Tombstone	A
40-105	Console	A
40-110	Tombstone	A
40-110	Console	A
40-115C	Table model	B
40-120C	Table model	C
40-120CI	Table model, ivory	A
40-124C	Table model	B
40-125C	Table model	C
40-130T	Table model	B
40-135T	Table model	B
40-140T	Table model (early)	B
40-140T	Table model (late)	B
40-145T	Table model	B
40-150T	Table model	C
40-155T	Table model	C
40-158F	Console	B
40-160F	Console	B
40-165F	Console	B
40-170CS	Chairside	A
40-180XF	Console	C
40-185XX	Console	A
40-190XF	Console	A
40-195XX	Console	B
40-200XX	Console	A
40-201XX	Console	A
40-205RX	Console	A
40-215RX	Console	A
40-216RX	Console	A
40-217RX	Console	A
40-501P	Table model radio-phono	A
40-502P	Table model radio-phono	A
40-503P	Table model radio-phono	A
40-504P	Portable radio-phono	A
40-506P	Console radio-phono	A
40-507P	Console radio-phono	A
40-508P	Console radio-phono	A
40-509P	Console radio-phono	A
40-510P	Console radio-phono	A
40-515P	Console radio-phono, walnut	A
40-515P	Console radio-phono, mahogany	A
40-516P	Console radio-phono	A
40-525P	Console radio-phono	A
40-526P	Console radio-phono	A
40-527P	Console radio-phono	A
PT-10C	Table model, various colors	A
PT-25	Table model	B
PT-26	Table model	B
PT-27	Table model	B
PT-28	Table model	A
PT-29	Tahle model	B
PT-31	Table model	B
PT-33	Table model	B
PT-35	Table model	A
PT-36	Table model	A

PT-37	Table model	A
PT-38	Table model	A
PT-39	Table model	A
PT-41	Table model	B
PT-43	Table model	A
PT-45	Table model	B
PT-46	Table model	A
PT-47	Table model	A
PT-48	Table model	A
PT-49	Table model	A
PT-50	Table model	A
PT-51	Table model	A
PT-53	Table model	A
PT-55	Table model	A
PT-57	Table model	A
PT-59	Table model	A
PT-61	Table model	A
PT-63	Portable	B
PT-65	Table model	B
PT-66	Table model	A
PT-67	Table model	A
PT-69	Table model w/clock	A
TH-14	Table model	A
TH-16	Table model	A
TH-15	Table model	A
TH-17	Table model	A
TH-18	Table model	A
TP-20	Table model	A
TP-21	Table model	A

1941

41-22CL	Table model w/clock	A
41-81T	Portable	A
41-83T	Portable	A
41-84T	Portable	A
41-85T	Portable	A
41-90CB	Table model	B
41-95T	Table model	B
41-100	Table model	A
41-100	Console	A
41-105	Table model	A
41-110K	Console	A
41-220C	Table model	C
41-221C	Table model	C
41-221CI	Table model	C
41-225C	Table model	B
41-226C	Table model	A
41-230T	Table model	C
41-231T	Table model	A
41-235T	Table model	C
41-240T	Table model	C
41-245T	Table model	C
41-246T	Table model	C
41-250T	Table model	C
41-255T	Table model	C
41-256T	Table model	B
41-258F	Console	A
41-260F	Console	A
41-265K	Console	B
41-280X	Console	C
41-285X	Console	B
41-287X	Console	A
41-290X	Console	A
41-295X	Console	B
41-296X	Console	B
41-300X	Console	B
41-315X	Console	A
41-316RX	Console	A
41-601P	Table model radio-phono	A
41-602P	Table model radio-phono	A
41-603P	Table model radio-phono	A
41-604P	Radio-phono	A
41-605P	Console radio-phono	A
41-607	Console radio-phono	A
41-608P	Console radio-phono	C
41-609P	Console radio-phono	A
41-610P	Console radio-phono	A
41-611P	Console radio-phono	A
41-616P	Console radio-phono	A
41-620P	Table model phono	A
41-623P	Table model radio-phono	A
41-624P	Console radio-phono	A
41-625P	Console radio-phono	A
41-629P	Console radio-phono	A
41-695P	Radio-phono	A

41-714	Table model	A
41-722	Table model	A
41-758	Table model	A
41-788	Table model	A
41-841T	Portable	C
41-842T	Portable	C
41-843T	Portable	C
41-844T	Portable	B
41-851T	Portable	C
41-KR	Table model w/clock	A
41-RP1	Table model phono	B
41-RP2	Table model phono	B
41-RP6	End table phono	A
PT-2	Table model	C
PT-6	Table model	B
PT-12	Table model	A
PT-30	Table model	C
PT-42	Table model	A
PT-44	Table model	B
PT-49	Table model	A
PT-87	Portable	B
PT-89C	Portable	A

1942

42-22CL	Table model w/clock	B
42-121CB	Table model	C
42-122T	Table model	C
42-123F	Console	B
42-124T	Table model	B
42-125K	Console	A
42-126T	Table model	A
42-321T	Table model	C
42-321TI	Table model	A
42-322T	Table model	C
42-323T	Table model	A
42-327	Table model	C
42-335T	Table model	B
42-340T	Table model	A
42-345T	Table model	B
42-350T	Table model	C
42-355T	Table model	B
42-358F	Console	B
42-360F	Console	B
42-365K	Console	B
42-380X	Console	C
42-390X	Console	C
42-395X	Console	B
42-400X	Console	A
42-620P	Table model phono	A
42-842T	Portable	B
42-843T	Portable	B
42-844T	Portable	B
42-853T	Portable	B
42-854T	Portable	B
42-1001	Table model radio-phono	B
42-1002	Table model radio-phono	A
42-1003	Table model radio-phono	A
42-1004	Console radio-phono	A
42-1005	Console radio-phono	A
42-1006	Console radio-phono	A
42-1008	Console radio-phono	A
42-1009	Console radio-phono	A
42-1010	Console radio-phono	A
42-1011	Console radio-phono	A
42-1012	Console radio-phono	A
42-1013	Console radio-phono	A
42-1015	Console radio-phono, walnut	A
42-1015	Console radio-phono, mahogany	A
42-1016	Console radio-phono	A
42-KR3	Table model	A
42-KR5	Table model	A
42-PT2	Table model	B
42-PT4	Table model	C
42-PT7	Table model	C
42-PT10	Table model	C
42-PT25	Table model	C
42-PT26	Table model	B
42-PT87	Portable	B
42-PT88	Portable	B
42-PT91	Table model	C
42-PT92	Table model	C
42-PT93	Table model	A
42-PT94	Table model	C
42-PT95	Table model	C
A-801	Chairside	C